# COLLINS HANDGUIDE TO THE
# BUT
# AND

## OF BRITAIN AND EUROPE

Painted by **John Wilkinson**

Text by **Michael Tweedie**

**Collins** St James's Place, London

William Collins Sons & Co Ltd
London · Glasgow · Sydney · Auckland
Toronto · Johannesburg

First published 1980

© in the text Michael Tweedie 1980

© in the illustrations John Wilkinson 1980

ISBN Hardback edition 0 00 219724 3

ISBN Paperback edition 0 00 219770 7

Colour reproduction by Adroit Photo-Litho Ltd, Birmingham

Filmset by Jolly & Barber Ltd, Rugby

Printed and bound by
William Collins Sons & Co Ltd, Glasgow

# Contents

The butterflies and moths shown here are species typical of the groups in which these insects are classified, and they show the order in which they are presented in the main text. The division into moths and butterflies is traditional and convenient rather than scientific; the great majority of the order Lepidoptera are regarded as moths.

Purple
Emperor 30–1

White
Admiral 31

Small
Tortoiseshell 33

Peacock 33

Red Admiral 34

Map 36

Fritillaries 37–43

4

Marbled White 44

Graylings 44–7

Mountain Ringlets 48–9

Gatekeeper 50–1

Wall Brown 54

Duke of Burgundy 55

Hairstreaks 56–7

Coppers 58–9

Blues 60–5

Skippers 66–9

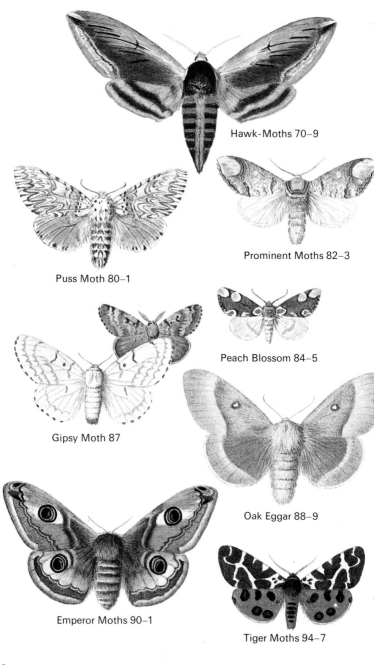

Hawk-Moths 70–9

Prominent Moths 82–3

Puss Moth 80–1

Peach Blossom 84–5

Gipsy Moth 87

Oak Eggar 88–9

Emperor Moths 90–1

Tiger Moths 94–7

6

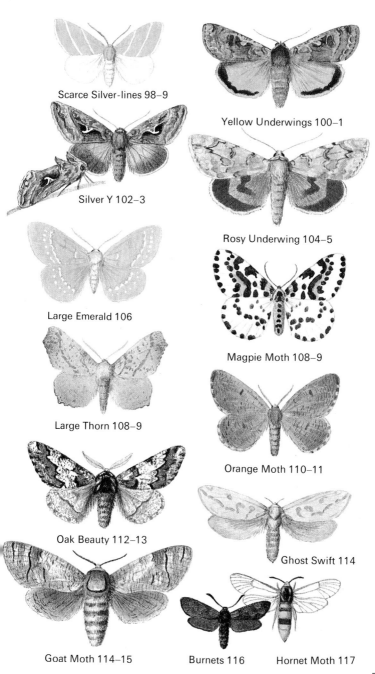

Scarce Silver-lines 98–9

Yellow Underwings 100–1

Silver Y 102–3

Rosy Underwing 104–5

Large Emerald 106

Magpie Moth 108–9

Large Thorn 108–9

Orange Moth 110–11

Oak Beauty 112–13

Ghost Swift 114

Goat Moth 114–15

Burnets 116

Hornet Moth 117

# Introduction

Butterflies and moths make up the order of insects called the Lepidoptera, a word meaning 'scale-wing', and it indicates their most important characteristic. The wing of a butterfly is covered by thousands of tiny overlapping scales, varying in colour but each one self coloured. These scales determine the pattern on the wing in the same way as the little pieces or tesserae make up the picture on a mosaic. The scales are loosely attached and readily come off if the insect is roughly handled. The wing-tip of the Emperor Moth (2) on the opposite page is enlarged to show the scales, and some greatly magnified scales are shown close to it.

One of the most frequent questions asked about butterflies and moths is: 'what is the difference between them?' This is best answered by explaining that the two together make up the Lepidoptera, which are classified in about 80 families. Of these the butterflies account for six and the far more diverse and numerous moths comprise the remainder; the distinction between them represents no important division in the natural classification. Butterflies are day fliers almost without exception: most moths fly by night but some are active by day. All butterflies have the antennae thickened or knobbed at the tip, and most of them sit at rest with the wings held together over the back. Both these characteristics crop up occasionally among the moths. A typical butterfly and moth, the Small Tortoiseshell (*Aglais urticae*) and the Emperor (*Saturnia pavonia*) are illustrated on the opposite page.

The skipper butterflies or Hesperioidea (pp. 66–9) exist in a sort of limbo. They are not at all closely related to the rest of the butterflies and are excluded by some authors of butterfly books, but no one ventures to call them moths. Since this book includes both categories of Lepidoptera they need not be assigned to either.

The division in English speech into butterflies and moths is inconvenient; it would be better if we had a word embracing all the Lepidoptera as the other European languages do. Butterflies are creatures of little importance and have never played much part in international commerce, either of goods or ideas. It is perhaps for this reason that the different languages have words for them which are entirely their own, showing no signs of common derivation. The French for butterfly (or moth) is *papillion*, which is indeed derived from the Latin *papilio*; but the Spanish is *mariposa* and the Italian *farfalla*. In German there are two words, *falter* and *schmetterling*, and the Dutch is *vlinder*. The Danes call butterflies *somerfugle* or 'summer-birds', but in Swedish they are *fjarilar*. When the distinction between butterfly and moth is needed the words for day and night are appended, as in the German *tagfalter* and *nachtfalter*.

The English word 'butterfly' is believed to have been originally applied to the Brimstone (*Gonepteryx rhamni*). The earlier inhabitants of Britain did not take much notice of butterflies, but they were always glad to see

**1** The Small Tortoiseshell butterfly (*Aglais urticae*).
**2** The Emperor Moth (*Saturnia pavonia*) showing a magnified wing-tip and the scales of which all Lepidoptera wings are composed.

any sign of the end of winter. Hibernated Brimstone butterflies sometimes appear as early as March, and the people called the lovely and welcome insect 'butterfly' from the yellow colour of its wings.

Europe and its offshore islands do not constitute a continent in any geographical sense; it is simply the western end of the great Eurasiatic continent, the cold and temperate parts of which are regarded by zoologists as comprising a region, the Palaearctic. The reality of this concept becomes apparent when we realise that of Britain's 70-odd species of butterflies over 20 (and many more moths) are common to this country and Japan. On the other hand, though none of the butterflies or larger moths are endemic to Britain, quite a number are found in Europe and nowhere else. A large proportion of these endemic species exist on the higher slopes of the Alps, Pyrenees and other mountains. This has come about because they are unable to exist in the lowlands, and so have become isolated on certain mountain ranges or parts of them. Isolation for long periods tends to lead to the evolution of distinct species. The Iberian Peninsula, isolated geographically from the rest of Eurasia by the Pyrenees, has a number of endemic moths and butterflies at all levels.

## Reproduction and Life History

All butterflies and moths lay eggs and the life history proceeds in three successive stages involving profound changes of form that are called metamorphosis. In the Lepidoptera the first stage after hatching, known as the larva or caterpillar, is that during which the insect feeds and grows. The second is the pupa or chrysalis, which is a resting stage and also serves to bridge the gap between the form of the larva and of the winged adult. The third stage, when the insect becomes winged and sexually mature, is known as the imago, plural imagines. On the opposite page the three stages of the White Admiral butterfly (*Limenitis camilla*) are shown (1).

The eggs of butterflies and moths are hard-shelled and although minute show considerable variation in form. The four examples shown opposite (3) are the eggs of the Orange-tip (*Anthocharis cardamines*), the Small Tortoiseshell (*Aglais urticae*), the Wall Brown (*Lasiommata megera*) and the Small Copper (*Lycaena phlaeas*). As a rule closely related species lay similar eggs: the flask-shaped form of the first is typical of the whites or Pieridae (pp. 24–9), the family to which the Orange-tip belongs.

When the young caterpillar is ready to hatch it bites a hole in the egg-shell and crawls out. The larvae of some species of moth invariably eat the shell after hatching, and will die if prevented from doing so. The staple diet of most butterfly and moth larvae is the leaves, buds or flowers of plants. Some, known as polyphagous larvae, will feed on a wide variety of plants, but most are more restricted in their diet: the caterpillar of the White Admiral feeds only on honeysuckle, those of the common whites of the genus *Pieris* only on plants of the family Cruciferae, which includes cabbage. A few feed in exceptional ways: the larvae of the Goat and Leopard Moths (p. 115) burrow in the trunks of trees, feeding on the wood, and that of the Large Blue (p. 60) spends part of its life in ants' nests devouring their larvae.

The eggs are almost always laid by the female imago on the leaves or stems of the larval food plant, stuck to them by an adhesive which quickly hardens when exposed to the air. She is informed of the correct food for her offspring by an inherited instinct and guided to the plant by the delicate sense of smell that resides in her antennae. On the opposite page a Small White (*Pieris rapae*) is shown laying an egg on a cabbage leaf (2).

The larva feeds voraciously and grows rapidly. In the course of doing so it sheds its outer skin at intervals, a feature of the growth of all insects. Butterfly larvae usually undergo this skin change or ecdysis four times before they change into a pupa, and with each ecdysis there is usually a change in appearance as well as in size. A larva about to change its skin spins a mat of silk on a leaf and firmly fastens its hindmost legs to it. It then waits for the skin to split behind the head and crawls out of the old skin, which is anchored by the hind legs to the mat of silk.

The caterpillar shown at the top of page 12 is that of the Black-veined White (*Aporia crataegi*). It consists of a small capsule-like head and a

1 The three stages – larva, pupa and imago – common to all Lepidoptera, in the life history of the White Admiral butterfly (*Limenitis camilla*).
2 A Small White butterfly (*Pieris rapae*) laying eggs on a cabbage leaf – the larval food plant of this species.
3 (from left to right) Eggs of the Orange-tip (*Anthocharis cardamines*), the Small Tortoiseshell (*Aglais urticae*), the Wall Brown (*Lasiommata megera*) and the Small Copper (*Lycaena phlaeas*).

flexible segmented body. The head is formed of a rigid substance called sclerotin and bears a pair of strong jaws for snipping or rasping off food. There is also a pair of minute antennae and a number, usually six, of very simple eyes called ocelli. With each ecdysis a new head capsule is formed behind the old one, which is discarded.

There are thirteen body segments of which the last two are fused together. The first three segments correspond to the thorax of the imago and bear the three pairs of jointed legs that are one of the basic characteristics of insects. Behind these are the abdominal segments of which the third to the sixth, and the last, each have a pair of prolegs or 'false legs'. These are fleshy and unjointed and each ends in a pad surrounded by a ring of minute hooks. In the large family of moths known as Geometridae (pp. 106–14) only two pairs of prolegs are present at the hind end of the body. Geometrid larvae walk by alternately extending and arching the body, and are often called looper caterpillars.

Some caterpillars are protected from predation by birds by repellant qualities such as spines or irritating hairs or an unpleasant taste. These are frequently conspicuous in appearance and gregarious, that is they live crowded together. The gregarious habit enhances their conspicuousness, which serves as a warning to birds that they are inedible. Those of the Peacock butterfly (*Inachis io*), shown at the bottom of this page feeding on nettle, are spiny and black and are very easy to see against the green foliage of their food plant.

When a butterfly larva is fully grown and ready to turn into a pupa, or pupate, it starts by hanging itself up. That of the Peacock spins a pad of silk on a plant stem and entangles the hooks of its hindmost prolegs in it, hanging head downwards. It remains in this position for a day or

**1**

more and then under-goes an ecdysis, but the discarded skin reveals not another larva but an object of totally different form, the pupa. The illustrations at the bottom of page 13 show four stages in pupation. On the left the skin has just split behind the caterpillar's head and in the next two illustrations it is seen being pushed upward over the body of the pupa. The last illustration shows the pupa hanging by its tail, having rid itself of the larval skin altogether. At the tail-tip of the pupa there is a number of minute hooks forming a structure called the cremaster. This is forced through the larval skin and entangled in the pad of silk, and the skin eventually shrivels and falls away. At first the pupa is soft and moist, but it soon dries and hardens and at the same time assumes its final shape.

**2**

In the swallowtail butterflies (pp. 22–3) and the whites and yellows (pp. 24–9) the pupa is suspended in a different way, head upwards and held by engagement of the cremaster in a pad of silk and also by a silken girdle round

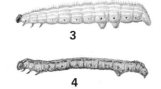

**3**

**4**

the middle of the body. It is supported in much the same way as a man at work on a telegraph pole.

The change in outward form from larva to pupa is remarkable enough, but the internal processes involved in the change are even more so. When the caterpillar hangs itself up to pupate, or if that of a moth spins its silken cocoon, a process called histolysis commences. In the course of this almost all its muscles and those of its internal organs that cannot serve the imago, are destroyed by wandering cells in the blood called phagocytes. These digest and liquify the tissues so that the greater part of the body cavity is filled with a sort of living soup. This substance provides the material and energy for building up the organs and appendages of the imago. These arise from clumps of cells called imaginal buds, which have resisted the process of dissolution and remained alive. In this way the substance of the larva is reconstituted to form a butterfly. The wings arise from inwardly growing imaginal buds and new legs,

**1** A Black-veined White caterpillar (*Aporia crataegi*) illustrating the small capsule-like head and flexible segmented body typical of the larval stage.
**2** Larvae of the Peacock butterfly (*Inachis io*) feeding on nettle.
**3, 4** Two examples of moth caterpillars which lack the normal five pairs of prolegs: the Silver Y (*Autographa gamma*) (**3**) with three prolegs and a member of the Geometridae, the Engrailed (*Ectropis bistortata*), which are characterised by only having two (**4**).
**5** Four stages in the process of pupation of the Peacock butterfly.

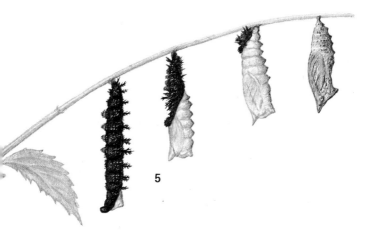

**5**

antennae and mouth-parts are formed at pupation and can be seen in outline on the shell of the pupa; formation of the adult internal organs continues after pupation.

Caterpillars spin silk to make a mat for anchorage when they change their skins and to give themselves a foothold for walking. In butterflies silk is used for suspension of the pupa, but in most moths the pupa is enclosed in a cocoon of silk, and that of the Domestic Silk Moth (*Bombyx mori*) is the basis of the silk industry. The silk is produced from a liquid contained in a gland in the labium or lower lip; when this is drawn out through a minute orifice it hardens and forms a fine filament. The pupae of Lepidoptera are of the type known as obtect, in which all the limbs and the external

1 Spurge Hawk Moth
2 Emperor Moth
3 Silver-washed Fritillary
4 Meadow Brown
5 Orange-tip
6 Grayling
7 Black-veined White

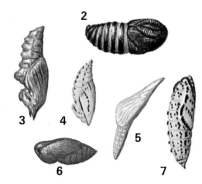

appendages are embedded in the shell. The pupa has therefore no power of controlled movement and must be anchored or enclosed. Most butterfly pupae are secured in the way already described, either hung head downwards or slung in an upright position by the tail and a girdle of silk.

In almost all moths the pupa is enclosed in a cocoon of silk. This may be pure silk like that of the Silk Moth (*Bombyx mori*), or mixed with earth or vegetable fragments. Larvae which burrow underground to pupate always construct an earthen cocoon, those which pupate above ground often spin two or more leaves together, or the pupa is concealed inside a hollow stem, as in some of the wainscot moths (p. 99). The larva of the Puss Moth (p. 80) makes a cocoon on the bark of the poplar or willow tree on which it has fed. By gluing together fragments of bark with liquid silk it constructs a shelter that protects the pupa in two ways. It has just the colour and texture of the bark and so is difficult to see, and it is also so hard that few enemies are able to penetrate it. It might be supposed that the moth would have difficulty in getting out, but at the time of emergence it secretes a solution of caustic potash which effectively softens the front of the cocoon. That of the Emperor Moth (p. 91) is strongly made and is pear-shaped with the narrow end in the form of a short tube which provides an exit for the imago. This exit is guarded by a ring of spines, pointing outwards and converging. From the inside they can easily be pushed aside, but they form an effective barrier against intruders.

Although moths comprise the great majority of the Lepidoptera and are far more diverse than the butterflies, the pupae of butterflies vary more in appearance than those of moths. Most of the latter are brown or blackish bullet-shaped objects which few entomologists can identify with confidence. That of the Spurge Hawk-moth (*Hyles euphorbiae*), shown at the top of page 14, is a typical example. Butterfly pupae are variously and specifically shaped and coloured (opposite page, figs. 3–7) and often effectively camouflaged in the surroundings in which the larva normally pupates. The explanation is of course that in almost all moths the pupa is concealed in a cocoon, and so natural selection does nothing to promote elaboration in their appearance, as this would serve no purpose. Butterfly pupae are almost always exposed and so benefit from any appearance that will conceal them from predators.

The final ecdysis is the appearance of the imago, and butterfly pupae usually change colour a day or so before this is due; the colour darkens and the pattern of the wings can often be seen through the pupa shell. At eclosion, as emergence is called, the pupa splits below the head and along the sutures marking the front edge of the wings, and a shield-shaped piece of the pupa is pushed off by the legs. The insect then crawls out and either remains clinging to the pupa shell or moves to a nearby position where its wings can hang downwards.

Three stages in the wing expansion of the Yellow-tail Moth (*Euproctis similis*) are shown on this page. At first the wings are like little crumpled bags, but liquid is pumped into them from the body and they expand. They do not swell up like balloons because the two membranes which form each wing are joined by strands which shorten as the wing enlarges. When they reach full size the fluid is withdrawn, the membranes come together and the wing flattens into a thin sheet supported by hollow rods; these are usually referred to as 'veins', but their chief function is to stiffen and support the wings. The insect must wait for an hour or two while the wings dry before it can fly. The whole process can be roughly simulated by crumpling up a paper bag, blowing it up and then flattening it out. While the wings are

**8** Three stages in the wing expansion of the Yellow-tail Moth (*Euproctis similis*).

drying the insect usually excretes a quantity of red fluid called meconium.

In temperate climates the life histories of butterflies and moths are always correlated with the seasons, and in many species there is a single complete cycle in the year. The winter is usually an inactive or hibernating phase, but this may correspond with any one of the four stages, egg, larva, pupa or imago. Larva and pupa are the stages which overwinter most frequently. A few species of moths fly during the winter months, among them the Winter Moth (p. 106). In this case it is only the male which flies, the female has only rudimentary wings. Frequently there are two cycles in the year, one of them being completed in the summer months; some small species breed continuously in the summer, completing three or more cycles in the year. On the other hand in northern regions low temperatures and short summers lead to the larvae of some species being unable to complete their growth in a single summer. Such larvae hibernate when partly grown, resume feeding after the winter and the imago appears after a total of two years of life. The Oak Eggar (p. 88) exists in two distinct races, a southern one in which the life history is completed in one year and a northern one, known as *callunae*, in which it takes two years.

The active life of the imago is usually a month or less, but species which overwinter in the adult state obviously live longer; for example a Peacock butterfly (p. 33), hatching from the pupa in August, may survive until the following May. A remarkable example of 'longevity' in Lepidoptera is recorded for the Small Eggar Moth (*Eriogaster lanestris*). It overwinters as a pupa, but the moths do not always appear the following spring; quite frequently eclosion is delayed for another year, and sometimes for even longer – a Small Eggar is known to have hatched seven years after pupation – but this is, of course, highly exceptional.

The sex of butterflies and moths is sometimes difficult to distinguish without close examination. Often they are quite distinct, as in the Gate-keeper (*Pyronia tithonus*), figures 2 and 3 opposite. Here the male has an oblique dark bar on the fore-wing that is lacking in the female. A more striking case of sexual dimorphism is seen in the Orange-tip (p. 27), where the orange colour is wholly absent in the female, and in the blue butterflies (pp. 60–5) the females are almost always heavily suffused with brown. Among the moths the Gipsy (p. 87) is a good example and in the Vapourer Moth (p. 86) the female is wingless and very heavy and stout. Here the sexes can be distinguished when the larvae are fully grown, the females being much larger than the males.

As many kinds of butterflies and moths fly together they need to be able to recognise members of their own species when seeking a mate. In butterflies both sight and scent play a part in this recognition; at a distance male and female react visually to each other's appearance, and it seems likely that much of the beauty and diversity of their coloration is to aid recognition; in addition, at close quarters, highly specific scented substances called pheromones enable them to be sure that each has selected the right partner for mating.

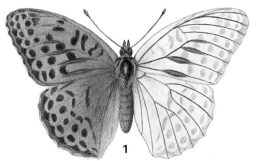

**1** A male Silver-washed Fritillary (*Argynnis paphia*), illustrated realistically and diagramatically to show the scent scales (androconia) which form streaks along the veins.
**2, 3** Male and female Gatekeeper butterflies (*Pyronia tithonus*) illustrating the distinctive dark bar of scent scales on the male's fore-wing.
**4** An enlarged head of a fritillary illustrating the long, clubbed antennae, the large compound eyes and the coiled proboscis characteristic of butterflies.

Many butterflies perform a courtship ritual before mating, and the female may be unwilling to accept the male until this has been completed. In these rituals an important part is played by special scented scales on the fore-wings of the male. These are called androconia, and in some species appear as black marks on the wings. In the Silver-washed Fritillary (*Argynnis paphia*) they take the form of streaks running along the veins (1), and in the Gatekeeper of a bar on the fore-wing (2). In courtship the androconia are scattered or otherwise applied to the female's antennae and she responds to their scent; they are in fact an aphrodisiac. Their scent can be detected by people with sensitive noses; that of the Wall Brown (p. 54) is said to smell like chocolate cream and that of the Green-veined White (p. 24) like lemon verbena.

Figure 4 shows the head of a butterfly, a fritillary, enlarged. The long clubbed antennae are segmented and the organs of smell are situated in minute pits on their surface. Below the base of the antennae are the large compound eyes: these do not form as clear images as ours do, but are extremely sensitive to movements. In front of them are the palps, equivalent to the labial palps of other insects and between and below these is the coiled proboscis, the organ by means of which the imago feeds. It is

Butterflies at buddleia bloom:
1 Peacock butterfly (*Inachis io*).
2 Meadow Brown (*Maniola jurtina*).
3 Small Skipper (*Thymelicus sylvestris*).

formed from the maxillae which are greatly prolonged and channelled along their inner surfaces. The two halves are joined together by minute hooks, rather like a zip fastener, so that the channels form a tube through which liquids can be sucked up. When not in use it is curled up like the hairspring of a watch, but can be quickly extended to probe a flower for nectar or to suck up fruit juice or moisture from the ground.

The first and part of the second legs seen in the drawing, illustrate a feature of the Nymphalid family of butterflies to which the fritillaries belong. In this family the first legs are small and stunted and are not used for walking; a Nymphalid butterfly rests on four legs like a quadruped. In some butterflies, including the Red Admiral (p. 34) the forelegs are known to be organs of taste and sensitive to very dilute sugar solutions.

Since the imagines of butterflies and moths mostly have short lives and do not grow, they need food mainly as a source of energy. By far the greater part of this consists of sugars from various sources, the most important being the nectar of flowers. Some species resort largely to gardens, especially the Vanessidi (pp. 32–5), and by growing the right flowers you can do much to encourage butterflies to come and enhance the beauty of your flower beds and shrubberies.

The best of the early butterfly flowers is certainly aubretia, a wall or rockery plant that likes chalky soils. It flowers in April and May and attracts both the hibernating species, the Peacock (p. 33), tortoiseshells (pp. 32–3) and Brimstone (p. 24) and the ones that hatch early like the Orange-tip (p. 27) and Green-veined White (p. 24). The flowering shrub buddleia is the finest butterfly lure of all and is sometimes called 'butterfly bush'. Lavender is a good butterfly plant and seems to be a favourite of the whites; while the ice plant (*Sedum spectabile*) will often be crowded with late summer butterflies if grown in a sunny position.

All the large daisy-like flowers are liked by butterflies for their plentiful nectar and for the firm support they give the insects to alight on. These include the single dahlias and chrysanthemums and above all the Michaelmas daisies which flower late, in September and October. They are specially beneficial to hibernating butterflies, the tortoiseshells, Peacock and Comma (p. 35), giving them a strengthening feast before they settle down for the winter. The two migratory species, Red Admiral and Painted Lady (p. 34) are believed to fly south from Britain to a warmer climate in the Autumn, and they too seek 'fuel' for their journey from the Michaelmas daisies.

Moths also visit flowers for nectar. Honeysuckle is a favourite of the long-tongued hawk-moths, and the beautiful Elephant Hawk (p. 79) can often be seen in June at dusk hovering in front of honeysuckle blossom; in much of Europe the Spurge Hawk (p. 77) is common and behaves in the same way. Later in the year you may be lucky enough to see the big Convolvulus Hawk (p. 74) feeding in the same way at honeysuckle or at the flowers of tobacco plant. This flower has a corolla tube so long that the Convolvulus Hawk is probably the only European moth that can reach its nectar. The Silver Y (p. 102) is often abundant in the summer and also hovers to feed at flowers: this is one of the very few moths that continue flying and feeding both by day and night. Many moths of the family Noctuidae (pp. 98–101) settle to feed on flowers. In the early spring the moths of that season come to the 'pussy willow' bloom of sallow bushes, and in autumn the ivy blossom is often thronged with noctuid moths.

**4**

**4** The pupa of the marsh-dwelling Swallowtail butterfly (*Papilio machaon britannicus*) attached to a reed.

A traditional method of collecting is that known as 'sugaring'. A sweet and preferably slightly fermented mixture is prepared and daubed on tree trunks and fence posts at dusk. At one time black treacle fortified with a little rum and stale beer was the fashionable recipe. Nowadays a less expensive compound of Barbados sugar and rotten bananas or apples is advocated. Whether you are a collector or not sugaring is great fun, but it is chancy. On a warm cloudy night you may see hundreds of moths feeding on and around your bait; on clear cool nights few or none will attend, and apparently favourable nights sometimes prove disappointing.

On the other hand many moths do not feed at all as adults, and the proboscis is rudimentary or absent. Among these are the eggars (p. 88), the emperors (pp. 90–1) and all the species that fly in the winter, when no food is available. In such cases enough food for their brief lives is carried over, stored in the body, from the larval stage.

## Where to Find Butterflies and Moths

Butterflies and moths can be found almost everywhere where there is some vegetation, from forests and marshes to gardens and city parks. Mixed woodlands are richest of all in species; the diversity of the Lepidoptera matches that of the plants on which they feed. All kinds of plants have butterflies and moths dependent on them for larval food, including even the mosses and lichens. Pine woods have their special moths, including the beautiful Spanish Moon Moth (p. 92), but the list of pine and fir feeders is not a very long one. Open moorland is the home of many species that are found in no other habitat, heather being the food plant of many of these. Numerous moths and some butterflies are restricted to marshes where a variety of species feed as larvae on reeds and rushes. Some live in the ordinary way on the leaves, others, such as the Bulrush Wainscot (p. 99) exist as larvae inside the stems feeding on the pith and pupating in the same situation. The British race of the Swallowtail (p. 21) is a marsh-dwelling butterfly; its pupa, attached to a reed, is illustrated on p. 19. Other species live in the sand dunes of the sea shore.

One of the most interesting of all habitats is that of the high mountain slopes between the highest trees and the summer snow. This is the home of most of the beautiful apollo butterflies (p. 21) and the mountain ringlets of the genus *Erebia* (pp. 48–9). Because mountain peaks and ranges are often isolated by intervening areas of lowland, many of them have species peculiar to them; this is especially true of the Erebias, some of which have a very limited distribution.

All these habitats, especially the lowland ones, are rapidly being destroyed and their butterflies and moths with them. They are beautiful creatures, and worth saving from extermination. But to do this sufficiently large areas of all their habitats must be saved from the greedy advance of concrete and the plough. It is hoped this book will encourage you to support every effort that is made to save them.

# The Scope of this Book

There are about 380 species of butterflies known to live in Europe of which less than 70 inhabit Britain; the moths number over two thousand. The selection of these shown in the following pages has been made to include firstly the common kinds that you are likely to see; secondly beautiful and striking species that you would like to identify if you do see them, and thirdly some butterflies and moths that are curious and interesting for one reason or another.

The classification follows the order used in most books on butterflies and moths, and the insects are shown against backgrounds that illustrate their usual habitat. Larvae of most of the species are included, resting on the plants that normally provide them with food. The species are identified by numbers correlated with the accompanying text, and the symbols ♂ and ♀ are used to indicate the male and female sex.

For those who wish to pursue the subject further there is an excellent book on the European butterflies: *A Field Guide to the Butterflies of Britain and Europe* by L. G. Higgins and N. D. Riley (Collins, London, 1970). There is no comprehensive book in English, at present in print, on European moths.

## PAPILIONIDAE

This family includes the swallowtails and some rather different butterflies such as the festoons and the apollos.

**1 Southern Festoon**, *Zerynthia polyxena*. Found in southern Europe and westward to Asia Minor, in the lowlands and up to 1,000m. It lives in rough stony places and its larva feeds on birthwort (*Aristolochia*).

**2 Apollo**, *Parnassius apollo*. A symbol of the high alpine meadows in which it lives, it is protected by law in much of Europe. In the Alps and Pyrenees it can be found up to 2,000m, but in Scandinavia it occurs in lowland country. The black, red-spotted larva feeds on species of stonecrop.

**3 Swallowtail**, *Papilio machaon britannicus*. The only member of the family found in Britain, where it is now confined to the Broads in Norfolk. British Swallowtails belong to a distinct subspecies which has stronger dark markings than the race found in France. Surprisingly the two subspecies live in very different habitats: *britannicus* is a marshland insect and its larva lives on milk parsley (*Peucedanum palustre*), while the continental race flies in open fields and downland with various umbelliferous plants as its larval food.

**4 Scarce Swallowtail**, *Iphiclides podalirius*. So called because it very occasionally strays to Britain. It is not uncommon in Europe and across temperate Asia to China. The larva feeds on species of *Prunus*.

**4**

♂

♀

1

♀

2

♂

♀

♂

3

♀

4

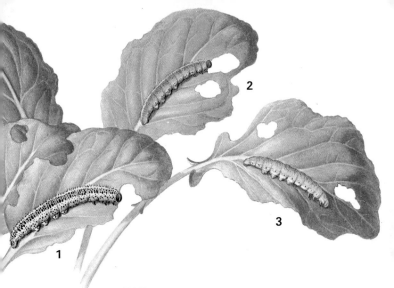

## WHITES AND YELLOWS

Butterflies of this family, also known as the Pieridae, are described on pages 24–9. All the species on this page are very common in Britain except for the Wood White which is local and rather rare. The first two, the Large White and the Small White, are often called 'cabbage whites' and they are the only European butterflies that have the status of pests.

**1 Large White**, *Pieris brassicae*. The male has the upperside fore-wings black-tipped but without any other black markings; the female has two black spots and a dark streak along the hind border of the wing. There are two broods in the year, the butterflies appearing in spring and summer; those of the latter brood are more heavily marked with black. The larvae feed on cabbage and also on 'nasturtium' (*Tropaeolum*).

**2 Small White**, *Pieris rapae*. Similarly marked and the sexes also distinguished by more dark markings on the fore-wing in the female. There is a spring and summer brood, the latter the more heavily marked of the two. The Small White is a serious pest of brassica crops and has been accidentally introduced into North America, Australia and New Zealand.

**3 Green-veined White**, *Pieris napi*. Very like the Small White but has distinctive dark markings along the veins of the wings. Its larval food is wild cruciferous plants and it does not attack cabbages.

**4 Wood White**, *Leptidea sinapis*. A curious little butterfly, very distinct from the other Whites and Yellows; most of its near relatives inhabit Central and South America. In Europe it is found locally in woods and the larva feeds on birdsfoot trefoil and other leguminous plants.

## WHITES AND YELLOWS

**1 Bath White**, *Pontia daplidice*. Received its English name because it was depicted in a piece of needlework by a young lady from Bath, in the late 18th century; the early entomologists were very whimsical in their naming of insects! The species is widely distributed in Europe and temperate Asia extending to Japan, but the British climate evidently does not suit it and its status in Britain is that of a rare migrant – 1947 was the last year in which it visited Britain in numbers. The larva feeds on various plants of the family Cruciferae.

**2 Orange-tip**, *Anthocharis cardamines*. Affords a striking case of sexual dimorphism: the broad orange areas on the wings of the male are absent in the female, and in flight she is not easy to distinguish from a Small or Green-veined White. The butterfly passes the winter as a pupa and appears in late April and May. The larva feeds on lady's smock, hedge mustard and other cruciferous plants and pupates in July, remaining as a pupa until the following spring. The pupa is curiously shaped, drawn out to a point in front and coloured like withered grass, and is well camouflaged among the dead herbage of the winter months. Several other kinds of Orange-tips are found in southern Europe. The male of the Morocco Orange-tip, *Anthocharis belia*, is a beautiful little butterfly with the ground of the wings bright yellow.

**3 Black-veined White**, *Aporia crataegi*. A common butterfly in Europe and North Africa and eastward to Japan. It existed in southern England until the 1920s but is now extinct. The larva feeds on hawthorn and various fruit trees and is sometimes a pest in orchards.

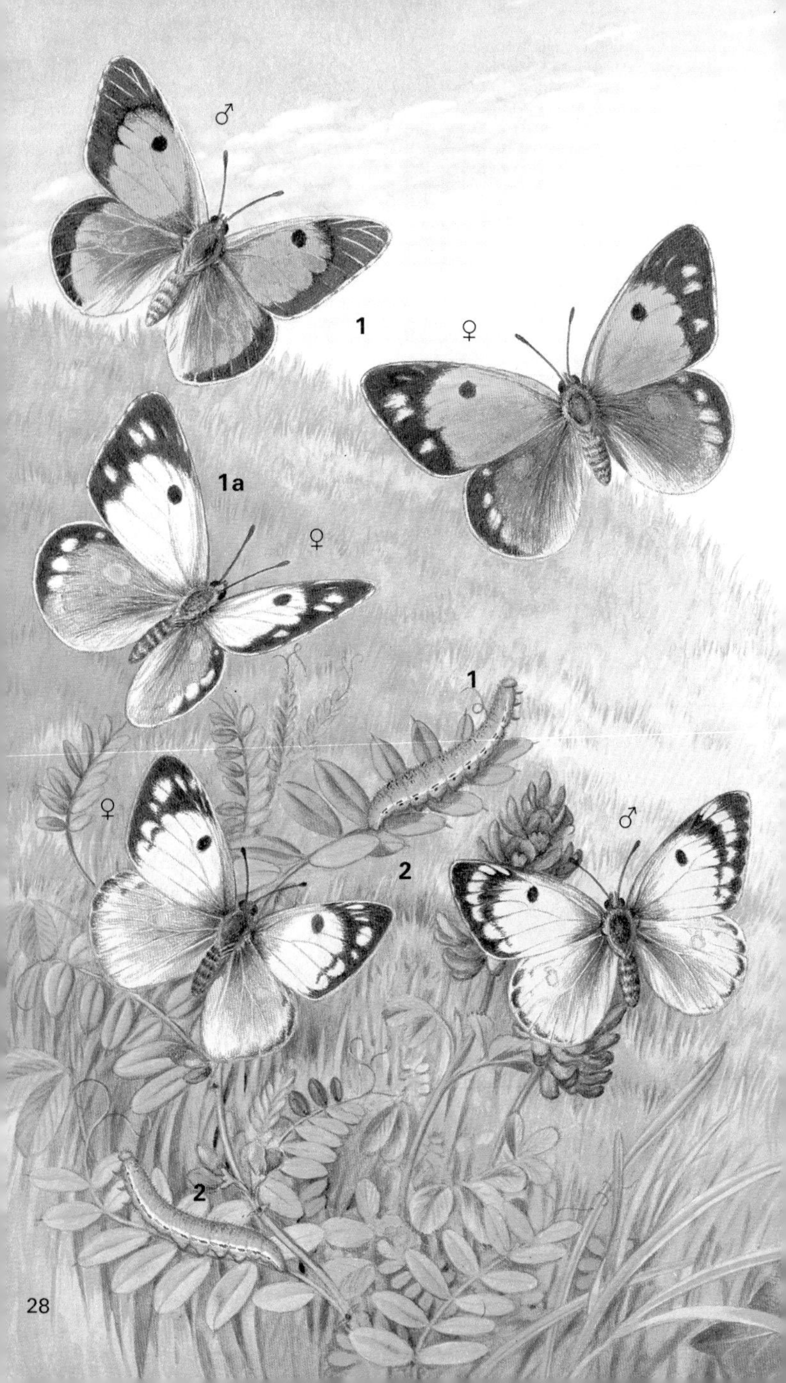
♂
1
♀
1a
♀
1
♀
2
♂
2

**1 Clouded Yellow**, *Colias crocea* and **2 Pale Clouded Yellow**, *Colias hyale*. Common in Europe but only migrants to Britain. They fly from the south in early summer and their descendants may appear in the autumn, but they cannot survive the winter in any stage. The Clouded Yellow is by far the more common of the two but is only abundant in certain favoured years, in which the Pale Clouded Yellow is generally also less rare than usual. Berger's Clouded Yellow, *Colias australis*, performs the same sort of migrations; this looks so like *Colias hyale* that the two were only recognised as distinct in 1947. The Clouded Yellow is typically deep orange in both sexes, the female being distinguished by having orange spots in the black borders of the wings. A variant form of the female known as *helice* (**1a**), with a creamy white ground colour, occurs in a ratio of about ten per cent in most populations and although it never appears in the male the gene that produces the pale colour is carried by both sexes. The Clouded Yellows frequent clover and lucerne fields and their larvae feed on plants of this family.

**3 Brimstone**, *Gonepteryx rhamni*. One of the species that hibernates through the winter in Europe as an imago. The male is coloured bright yellow, the female pale greenish-yellow. The butterflies mate in the spring and the eggs are laid on buckthorn. Both larva and pupa are green and are well camouflaged in the summer foliage.

## NYMPHALIDAE

The butterflies of this family (pp. 30–43) are usually large and brightly coloured and they all have the front pair of legs much reduced in size and not used for walking. These legs are brush-like in males but only have short bristles in females.

**1 Purple Emperor**, *Apatura iris*. A magnificent butterfly that lives in old-established woodlands, especially oak woods, in southern England and all over Europe and temperate Asia to Japan. Only the males display the purple sheen on the upperside of the wings; the larval food plant is sallow.

**2 Poplar Admiral**, *Limenitis populi*. Ranges from central Europe east-wards and not uncommon; it is absent from Britain, western France and Spain. Its larva feeds on aspen and other kinds of poplar.

**3 White Admiral**, *Limenitis camilla*. Absent from Spain and Portugal but is widespread in central Europe and is found in mixed woodland in the southern half of England, where it has extended its range and become more common since about 1920, but of course its continued existence depends on the preservation of woodland. The larva feeds on honeysuckle and hibernates when young in a shelter made by spinning the edges of a leaf together and securing it with silk to the stem.

**1 Camberwell Beauty**, *Nymphalis antiopa*. Widespread in the northern hemisphere and in North America is called the Mourning Cloak, a translation of the German *Trauermantel*. It is rarely seen in Britain though it is an occasional autumn migrant from Scandinavia, and it is possible that some butterflies are brought over on timber ships. However, the butterfly requires the sustained cold of a continental winter to hibernate successfully and in the damp mild British winter it fails to go into the deep sleep that is an essential part of its life cycle. The caterpillar feeds on forest trees including birch, willow and sallow. The name 'Camberwell Beauty' commemorates the fact that the first recorded British specimens were taken at Camberwell, now a part of urban London, but at that time a rural village.

**2 Large Tortoiseshell**, *Nymphalis polychloros*. Not uncommon in south-eastern England in the period between the two World Wars, but has been rarely seen in Britain since. The larvae live gregariously high up in elms, and sometimes willows and poplars, and it seems likely that the recent epidemic of Dutch Elm disease will reduce the Large Tortoiseshell's numbers still further. Like the other species on this page it is a winter hibernator. It is not uncommon in Europe and extends eastwards to the Himalayas.

**3 Small Tortoiseshell**,
*Aglais urticae.* Often spends
the winter in sheds and
attics and appears fluttering
inside windows in the first
warm days of spring. It is a
very common and
widespread butterfly whose
larva feeds on nettle.
Superficially it resembles the
Large Tortoiseshell but its
structure and anatomy place
the Small Tortoiseshell in
another genus, *Aglais.*

**4 Peacock**, *Inachis io*
Another common British
and European species that
often hibernates indoors and
flies on warm days in early
spring. Its larva, like that of
the Small Tortoiseshell, Red
Admiral and Comma, feeds
on nettle; so if nettles are
ruthlessly destroyed these
beautiful butterflies will
become rare or disappear.

**1 Red Admiral**, *Vanessa atalanta*. A common butterfly in Europe, North America and temperate Asia. In northern Europe it does not survive the winter in any stage, but every spring there is a northward migration from the Mediterranean region, and it is the offspring of these butterflies that we see on our autumn flowers. The larva feeds on nettle.

**2 Painted Lady**, *Cynthia cardui*. Probably the most widely distributed butterfly and known in North America by the appropriate name 'Cosmopolite'. It is found all round the world wherever butterflies can be expected to live, from the cold temperate regions to areas of semi-desert and at moderate altitudes on tropical mountains; it is least common in South America. In Australia it is represented by a very similar species, *Cynthia kershawi*, that is sometimes regarded as a subspecies of *cardui*. Its wide range is associated with the habit of migration, and it has the same status in Britain as the Red Admiral. The larva feeds on various kinds of thistles.

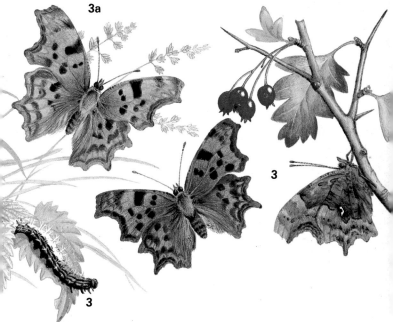

**3 Comma**, *Polygonia c-album*. One of the butterflies that hibernates as an adult in the open in Britain. It is most effectively camouflaged by the pattern of the underside and the irregular outline of the wings, which give it a remarkable resemblance to a dead crumpled leaf, the white 'comma' looking like a crack in the leaf. Eggs are laid in the spring and pale coloured *hutchinsoni* butterflies (**3a**) may appear as early as June and July; if the larvae feed up more slowly to produce late summer butterflies these are of the normal dark form. Nettle is the usual food plant but the larva also feeds on hop and currant.

## NYMPHALIDAE

**1 Map Butterfly**, *Araschnia levana*. Completes two life cycles in the year, the butterflies appearing in May and July. Slight differences in broods are seen in the whites (p. 24), but in this species the two broods could be mistaken for different species. The 18th-century naturalist Linnaeus, the first authority to name and classify animals systematically, did think they were two species, which he named *levana* and *prorsa*, and the second name is now used for the summer form (**1a**) which looks like a small White Admiral (p. 31). The food plant is nettle and the eggs are laid one on top of the other, forming a column a dozen or so eggs high. The Map is not found in Britain but extends into north-eastern France.

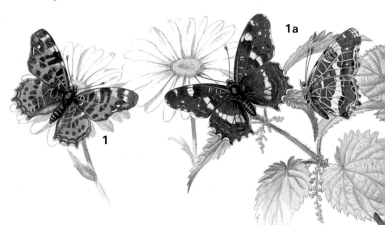

*Fritillaries*

The fritillaries (pp. 36–43) share their name with a well-known flower, and the reference in both cases is to the Latin *fritillaria* a checker-board. They fall conveniently into two groups, those with silvery-metallic markings on the underside (pp. 36–41) and those without (pp. 42–3). The larvae of the first group practically all feed on violet plants.

**2 Silver-washed Fritillary**, *Argynnis paphia*. The metallic markings take the form of ill-defined stripes, and the sexes differ by the presence of black streaks of scent-scales on the fore-wing of the male (p. 17). A not uncommon form of the female, known as *valesina*, has the ground colour greenish-grey. It does not appear in the male, though the gene for it is carried by both sexes. The case is analogous to that of form *helice* of the Clouded Yellow (p. 29). This is a woodland butterfly, quite common in Europe including Britain.

**3 High Brown Fritillary**, *Fabriciana adippe*. The silvery markings take the form of rounded spots and it is rather smaller than *A. paphia*. It is also a woodland species and is found in England north to the Lake District and widely in Europe.

2

♀

♂

2

3

3

**1 Dark Green Fritillary**, *Mesoacidalia aglaja*. Very like the High Brown but the silver spot spots on the hind-wing underside are not surrounded with reddish-brown and the basal half of the wing is dull green. The butterfly inhabits more open country, chalk downs moorland and coastal sandhills. In Britain it ranges much further north than the High Brown, reaching the Western Isles and the Orkneys and extending to Ireland.

**2 Cardinal**, *Pandorina pandora*. The largest and finest of the European fritillaries. It resembles the Silver-washed Fritillary but is distinctly larger and the central area of the fore-wing underside is rose-red. It extends from southern Europe eastward to Iran and is found in flowery meadows from sea level to about 1,200m.

**3 Queen of Spain Fritillary**, *Issoria lathonia*. The most beautiful of the 'silver-spot' fritillaries. The metallic spots on the underside are larger and brighter than in the other European species and the fore-wings more pointed in shape. It is an inhabitant of southern Europe and North Africa, whence it performs migratory flights northwards, occasionally reaching southern England. In 1872, a great year for migratory butterflies, about 50 were recorded, but years may go without it being seen in Britain.

**3**

**4 Lesser Marbled Fritillary**, *Brenthis ino*. A small fritillary without metallic markings on the underside. It is found in damp marshy meadows and valleys in central and northern Europe up to Lapland though not in the British Isles or north-western France. The larval food plants are meadow-sweet, raspberry and great burnet. The two other species of *Brenthis*, *B. hecate* and *B. daphne*, have a southern distribution.

**4**

**4**

## NYMPHALIDAE

The little fritillaries of the genus *Clossiana* are related to the larger ones described on pages 36–8. They have silvery spots on the hind-wing underside and in latitudes where species of violet are available this is the larval food plant. A few of the species are found on the tundra of the far north. The Arctic Fritillary, *C. chariclea* has been taken, together with the Small Copper (p. 58) in Ellesmere Island in Canada between 81° and 82°, probably farther north than any other butterflies. The three illustrated here are all found in temperate regions. The Violet Fritillary does not extend to Britain but the other two are quite common in this country.

**1 Pearl-bordered Fritillary**, *Clossiana euphrosyne.* Appears in woodlands in May and is often abundant in the southern half of Britain, becoming rarer northwards. In an early 18th-century book on butterflies the Pearl-bordered was named the April Fritillary, but it very seldom flies as early as this. Professor E. B. Ford has suggested that before the revision of the calendar in 1752, when eleven days were omitted from the year, late April would have been the time of its appearance.

**2 Small Pearl-bordered Fritillary**, *Clossiana selene.* Distinguished from *euphrosyne* most easily by the pattern of the hind-wing underside which has darker brown markings and a larger number of metallic spots. The Small Pearl-bordered flies a little later, emerging in June, and frequents damp woodlands and marshy open country as well. It replaces the Pearl-bordered as the common small fritillary in northern Britain, though the two fly together over a great part of the country. Only *euphrosyne* extends to Ireland, and it is restricted to one locality there. Both extend from Europe to far-eastern temperate Asia.

**3 Violet Fritillary**, *Clossiana dia.* A more southern species, not extending further north than France and Germany. It is darker and more heavily marked than the other two species.

2                                                                        1

1

1

2

4 ♂

4

2

3

3

4 ♀

The three genera *Melitea*, *Mellicta* and *Euphydryas* comprise the fritillaries without metallic markings on the underside. There are numerous species in Europe and temperate Asia, some lowland butterflies, others living high in the mountains. Some of the species are difficult to distinguish from each other, chiefly because most of them have variable wing patterns. That of the hind-wing underside usually has the most reliable features, and on the right side of this page the underside of each of the four species is shown. Most larvae are known to feed on low-growing plants, plantain being favoured by several species. In the British Isles they are poorly represented, only one of each genus being present.

**1 Spotted Fritillary**, *Melitea didyma*. Though not British, this is a widely distributed butterfly ranging east to central Asia and from the lowlands up to 2,000m. The butterfly varies regionally, and the two or three annual broods even tend to differ from each other in the same locality. However its continuous distribution makes it impossible to separate definite geographical subspecies.

**2 Glanville Fritillary**, *Melitea cinxia*. Derives its English name from that of a lady who collected butterflies in the early 18th century. It is widespread in Europe in the lowlands and fairly high in the mountains, but in Britain it is on the extreme northern limit of its range. At one time it existed at several localities on the south coast but is now confined to the Isle of Wight.

**3 Heath Fritillary**, *Melicta athalia*. Also a very variable butterfly and the commonest member of the genus. It ranges from western Europe to Japan and is found in a few widely separated localities in southern England. The most favoured food plant is cow-wheat.

**4 Marsh Fritillary**, *Euphydryas aurinia*. A variable species that often exists in colonies which acquire characteristics of their own, and there is much individual variation as well. It is found in Great Britain and Ireland and frequents marshes, moors and chalk downs where its food plant devil's-bit scabious grows; it extends right across Europe and Asia.

## BROWNS

The family characteristics of the Browns are described on page 47.

**1 Marbled White**, *Melanargia galathea*. The most common of a number of similar species found in Europe and western Asia. It is widely distributed in Europe, including England; the dark form *procida* (**1a**) occurs in southern Italy and the Balkans. The larva feeds on grasses.

**2 Grayling**, *Hipparchia semele*. One of a number of European species of this genus all rather similar in appearance and often difficult to identify. *H. semele* is the only one that is found in Britain, and it extends all over the British Isles including the Hebrides, inhabiting dry situations near the coast, and moorland

and chalk downs inland. The hind-wing underside pattern gives very effective camouflage against the ground in such places and also against the bark of trees. When the butterfly settles, it waits for a moment and then slips its fore-wings down between the hind-wings, concealing the eye-mark on the fore-wing, and leaning to one side to avoid casting a shadow. In courtship between the sexes, the male transfers his scent scales to the female by clasping her antennae between his wings. All graylings feed as larvae on grasses.

**3 Tree Grayling**, *Hipparchia statilinus*. Much darker in colour than the Grayling. The male is very dark brown on the upperside with the eye-like spots hardly visible in some specimens. It lives in central and southern Europe.

**4 Southern Grayling**, *Hipparchia aristeus*. One of the most distinctive species, having both fore- and hind-wings broadly marked with orange-brown. It is found only in the countries bordering the Mediterranean.

The butterflies of the family Satyridae (pp. 44–55) are commonly known as 'browns' and most of the species are brown with the underside more distinctly patterned than the upperside; the marbled whites (p. 44) are one of the exceptions to this rule. The wing pattern often consists largely of pupilled rings or 'eye-spots' and it is believed that they act as deflection marks and afford protection against attack by birds, which peck at them because they are a conspicuous feature and do no more than take a small piece out of the wing. Browns feed as larvae on monocotyledenous plants, grasses in temperate climates, bamboos and palms in the tropics. None of the three species described on this page is found in Britain.

**1 Dryad**, *Minois dryas*. A dusky brown butterfly distinguished by the blue pupils of the eye-spots on the fore-wing. It flies in lightly wooded places and is local in its distribution in central Europe.

**2 Rock Grayling**, *Hipparchia alcyone*. Inhabits steep rocky country, where it flies from the foothills up to 2,000m. It extends from central and south Europe to Asia Minor.

**3 Great Banded Grayling**, *Brintesia circe*. The finest of all the European Browns. The female is larger than the male but otherwise the sexes are similar, a rather unusual feature in Satyrids. It is found mainly in lightly wooded lowlands and foothills and extends from south and central Europe east to the Himalayas.

3

3

3

1

1a

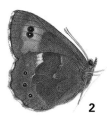

2

3

4

### Mountain Ringlets

The mountain ringlets of the genus *Erebia* (pp. 48–9) form a very large genus of several hundred species, mostly from the high mountains or the northern regions of Europe and Asia, with a few in North America. Together with the apollos (*Parnassius*, p. 23) they are the characteristic mountain butterflies, though some occur in the lowlands, especially in the more northern countries. The large number of species is probably due to their mountain habitat combined with rather feeble powers of flight so that populations have become isolated and evolved into distinct species. Many resemble each other and are hard to identify.

**1 Mountain Ringlet**, *Erebia epiphron*. Numerous local subspecies two of which are shown here. The typical form *E. e. epiphron* (**1a**) is found in the mountains of Czechoslovakia and formerly existed in the Hartz Mountains. In the Alps the species is represented by *E. e. aetheria* (**1**) where it lives on the slopes above 1,700m. A third subspecies, *E. e. mnemon* is found in the mountains of Scotland and northern England. In sunny weather the butterflies fly close to the ground but if a cloud comes over the sun they disappear into the grass and are extremely hard to find.

**2 Arran Brown**, *Erebia ligea*. Occurs in central and northern Europe, in the lowlands in the more northern regions. It is just possible that this species lives in Scotland and the Western Isles, but this is still uncertain. It is distinguished from the very similar Scotch Argus by a white spot on the margin of the underside hind-wing.

**3 Scotch Argus**, *Erebia aethiops*. Widespread in Europe and east to the Caucasus and Ural Mountains. It occurs up to 2,000m, but is not a mountain butterfly, and flies over damp lowland moors in Scotland and other northern parts of its range; in Belgium it is found on coastal sand dunes. Like the Mountain Ringlet this species will only fly if the sun is shining.

**4 Woodland Ringlet**, *Erebia medusa*. Another frequenter of moorland and damp meadows from the lowlands to 1,200m. It is found in central and eastern Europe.

## BROWNS

**1 Meadow Brown**, *Maniola jurtina*. One of the most abundant European butterflies and is found all over the British Isles. In north-western Scotland, the Western Isles, Ireland and the Isles of Scilly the Meadow Brown is represented by a more brightly coloured form with the orange markings more extensive in both sexes. These are all included in the subspecies *M. j. splendida*, but each population has distinctive features. In south-western Europe and north-western Africa there is another large and brightly coloured subspecies, *M. j. hispulla*. The Meadow Brown flies in a variety of habitats, meadows, roadsides, borders of woods and gardens. The larva feeds on various grasses but is seldom seen as it hides at the grass roots by day, coming out to feed only at night.

**2 Gatekeeper**, *Pyronia tithonus*. Sometimes called the Hedge Brown. It is another very common butterfly, most often seen along hedges where blackberry or bramble is in flower. The name 'Gatekeeper' may refer to its frequent occurrence near field gates, or perhaps to the keeper of a toll gate, who was a man who spent his life beside the road 150 years ago, when British roads were more quiet and flowery than they are now. Even today the butterfly is often seen along country lanes. The male has a conspicuous band of black scent scales on the fore-wing, and is illustrated on page 17. In Europe the species extends from Spain to Asia and the Caucasus. Two other very similar species of *Pyronia* are found in southern Europe, the Southern and the Spanish Gatekeeper, *P. cecilia* and *P. bathseba*.

**3 Ringlet**, *Aphantopus hyperantus*. The underside is pale brown with conspicuous eye-spots or ocelli. These are very variable in their development. Not uncommonly the yellow rings of the ocelli become reduced in diameter or absent, only the pupils remain as small white dots; this is the form known as *caeca*. Conversely the rings may be enlarged and distorted into ovals; a form, known as *lanceolata*. The butterfly favours damp situations, both meadows and open paths and clearings in woods. It is confined to the central and northern parts of Europe, extending east into Asia, but is absent from the Mediterranean region, Italy, Greece and most of Spain. In the British Isles it is absent only from northern Scotland and the Western Isles.

## BROWNS

**1 Large Heath**, *Coenonympha tullia*. A very widespread and variable butterfly that extends all round the cool-temperate parts of the northern hemisphere. Variation is expressed mainly in the development of the eye-spots or ocelli on the hind-wing underside; and it is to a great extent regional. In the central European form *C. t. tiphon* (**1a**) they are moderately well developed. In Europe the various subspecies fall into two groups, separated regionally and ecologically. In north, central and western Europe the butterfly inhabits damp moorland from the lowlands to moderate altitudes. In Italy and the Balkan countries the Large Heath is definitely a mountain butterfly. In the British Isles no less than three subspecies are found: in the Scottish *C. t. scotica* (**1**) the underside ocelli are very small or absent; the typical form *C. t. tullia* is found in south Scotland and the most northern parts of England, with the ocelli moderately enlarged; while in north-western England a form called *C. t. rothliebii* occurs

having the ocelli very large and prominent. These three subspecies merge into one another to form a continuous series, which is known as a cline. In the Shetland islands the Large Heath (form *scotica*) is the only indigenous butterfly.

**2 Pearly Heath**, *Coenonympha arcania*. One of the prettiest species in the genus. The underside of the hind-wing is beautifully patterned with a white bar and yellow-ringed white ocelli. It is found in almost all of south and central Europe right up to the Channel coast, but not in Britain.

**3 Small Heath**, *Coenonympha pamphilus*. One of the more widespread species, being found all over Europe and the British Isles eastward into Russia. It is not subject to much variation. The genus *Coenonympha* has eleven species in Europe some of which have a restricted distribution. The Spanish Heath, *C. iphioides* is confined to Spain and the Corsican Heath, *C. corinna* to Corsica, Sardinia and Elba. Their appearance is fairly uniform with one exception, the False Ringlet *C. oedippus*, which looks remarkably like a small Ringlet (p. 51).

## BROWNS AND NEMEOBIIDAE

**1 Wall Brown**, *Lasiommata megera*. A sun-lover and believed to be named after its fondness for basking on walls in full sunshine with its wings outspread. The male and female butterflies are quite distinct, the former having an oblique band of scent scales on the fore-wing, which are said to smell like chocolate cream. The butterfly is common in Europe, including the British Isles, and extends eastward to Iran. In Britain it completes its life cycle twice in the year, flying in May and again in July and August.

**2 Large Wall Brown**, *Lasiommata maera*. Widespread in Europe and east to central Asia, but is not found in Britain. In Europe it appears in two forms: a northern one which is relatively dull in coloration; and one found in southern France, Spain and Portugal, called *adrasta*, that has the fulvous-orange colour much more extensive. The picture on this page is of a female of the form *adrasta*.

**3 Speckled Wood**, *Pararge aegeria*. As much a lover of woodland shade as the Wall Brown is of the sun. It occurs in Europe in two very distinct forms: the northern form, which is shown here, is *P. a. tircis* and inhabits northern, central and eastern Europe and the British Isles; the southern form, *P. a. aegeria*, has dull orange instead of creamy-white markings and flies in southern Europe. In Britain the Speckled Wood may pass the winter as a larva or a pupa. Butterflies from the latter appear in early April, those from the former in May. Second broods appear later in the summer.

**4 Duke of Burgundy**, *Hamearis lucina*. The only European species of the family Nemeobiidae which is best represented in the American tropics. It is well distributed in Europe and extends to southern England. The usual habitat is woodland clearings and the larva feeds on primrose and cowslip.

## LYCAENIDAE

This enormous butterfly family (pp. 56–65) is well represented in Europe where it falls into three groups, the hairstreaks, coppers and blues.

### Hairstreaks

These butterflies (pp. 56–7) are mostly dark on the upperside, variously coloured on the underside with a line running across both wings, the 'hairstreak'. The larvae of nearly all of them feed on trees and bushes.

**1 White-letter Hairstreak,** *Strymonidia w-album*. Owes both its English and Latin names to the white W-mark on the hind-wing underside. In England it is only found in the southern half, in Spain in the north, but otherwise it occurs all over Europe and eastward to Japan. The larva feeds on elm, especially wych elm, and the imago visits bramble flowers.

**2 Brown Hairstreak**, *Thecla betulae*. The largest of the European hair-streaks. The sexes can be readily identified by the orange patch on the fore-wing which is well developed in the female but obscure in the male. The species lives in open woodland and the most usual food plant is sloe. Both this and the last species overwinter as eggs. Brown Hairstreaks occur over most of Europe, in England and in the south of Ireland.

**3 Green Hairstreak**, *Callophrys rubi*. The underside is green with the 'hairstreak' reduced to a row of white dots or absent. The larva feeds on gorse and broom and the butterfly may occur wherever these plants grow, from the lowlands to 2,300m. It is found all over Europe and Britain.

**4 Purple Hairstreak**, *Quercusia quercus*. Found all over Europe and in the British Isles. The female has a patch of rich dark blue in the basal area of the fore-wing; in the male both wings are glossed all over with purplish-blue. The larva feeds on oak and the butterflies are seen flying around the outer branches of oak trees.

**5 Ilex Hairstreak**, *Nordmannia ilicis*. Inhabits most of Europe, though not Britain, and extends east to Asia Minor. It flies on rocky slopes where small shrubby species of oak are growing and the larva feeds on these.

## LYCAENIDAE

### *Coppers*

The coppers (pp. 58–9) include some of the most beautiful of the Lycaenidae and are mainly found in the Eurasian and North American continents. In most of the species the females are more heavily marked on the upperside than the males. The larvae feed on low-growing plants, dock being the most usual food.

**1 Scarce Copper**, *Heodes virgaureae.* So named because it was once believed to occur as a rarity in England. The male has the upperside brilliant red-gold, usually without any markings; the female is quite heavily marked with black spots. The species is common locally in central Europe and extends eastwards into Asia.

**2 Purple-shot Copper**, *Heodes alciphron.* Varies regionally in Europe. In the south, especially in the Alps, a brightly coloured subspecies occurs, *H. a. gordius* and it is the male of *gordius* that is shown here. In central Europe the typical form, *H. a. alciphron* is found, in which both sexes have the upperside heavily suffused with brown, the male with a violet gloss.

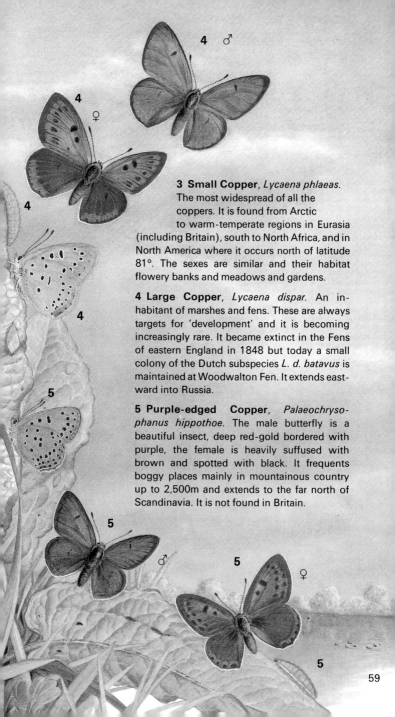

**3 Small Copper**, *Lycaena phlaeas*. The most widespread of all the coppers. It is found from Arctic to warm-temperate regions in Eurasia (including Britain), south to North Africa, and in North America where it occurs north of latitude 81°. The sexes are similar and their habitat flowery banks and meadows and gardens.

**4 Large Copper**, *Lycaena dispar*. An inhabitant of marshes and fens. These are always targets for 'development' and it is becoming increasingly rare. It became extinct in the Fens of eastern England in 1848 but today a small colony of the Dutch subspecies *L. d. batavus* is maintained at Woodwalton Fen. It extends eastward into Russia.

**5 Purple-edged Copper**, *Palaeochryso-phanus hippothoe*. The male butterfly is a beautiful insect, deep red-gold bordered with purple, the female is heavily suffused with brown and spotted with black. It frequents boggy places mainly in mountainous country up to 2,500m and extends to the far north of Scandinavia. It is not found in Britain.

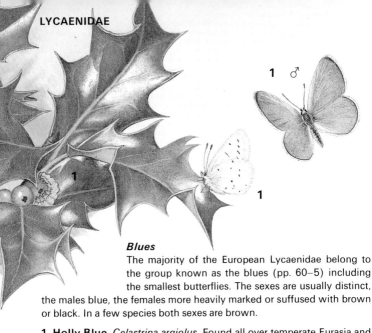

### Blues

The majority of the European Lycaenidae belong to the group known as the blues (pp. 60–5) including the smallest butterflies. The sexes are usually distinct, the males blue, the females more heavily marked or suffused with brown or black. In a few species both sexes are brown.

**1 Holly Blue**, *Celastrina argiolus*. Found all over temperate Eurasia and North America. The female differs from the male in having broad black borders to the wings. In Britain the butterfly has two annual broods: spring larvae feed on flowers of holly, autumn ones on ivy buds.

**2 Large Blue**, *Maculinea arion*. Lives in grassy places where thyme grows. The larva feeds on thyme for a short period but is then taken into their nests by ants of the genus *Myrmica* where it feeds on the ants' larvae and pupae (see also p. 64). The Large Blue is probably extinct in Britain.

**3 Long-tailed Blue**, *Lampides boeticus*. Almost worldwide in the tropics and warm-temperate regions; its habit of migration brings it occasionally to Britain. It flies in flowery meadows and gardens, and the larva feeds on various leguminous plants, living in the seed-pods.

**4 Short-tailed Blue**, *Everes argiades*. A rather small species in which the female is brown with little or no blue on the wings; the 'tail' on the hind-wing of both sexes is short and inconspicuous. This is a widespread butterfly that extends to Brittany and, it seems, might well be able to live in England, but it is seen there only very occasionally.

**5 Little Blue**, *Cupido minimus*. One of the smallest European butterflies. Both sexes are brown but only the male has scattered blue scales on the upperside. The larva feeds on small leguminous plants. Widespread in Europe, mainly on chalky soil, and extending to England and Ireland.

## LYCAENIDAE

**1 Green-underside Blue**, *Glaucopsyche alexis*. Distinguished by the greenish basal area of the hind-wing underside. It flies near woodland in hilly country over most of Europe, but is absent from Britain. The larva lives on various small leguminous plants.

**2 Brown Argus**, *Aricia agestis*. Inhabits heaths and downs all over Europe, including southern England, east to Siberia. The sexes are similar and neither has any blue coloration. The larva feeds on rock-rose and stork's bill.

**3 Mountain Argus**, *Aricia allous*. Very similar to the Brown Argus, but the orange lunules on the borders of the wings are less developed and often absent. It replaces the Brown Argus above 1,300m in central Europe and in the lowlands in northern regions, including northern England. In Scotland a very distinct form *artaxerxes* occurs with a white spot in the centre of the fore-wing.

**4 Silver-studded Blue**, *Plebejus argus*. A characteristic butterfly of heaths, and the larva feeds on flowers of gorse and broom and also heather and bilberry. It is found throughout Europe east to Japan and in southern England and Wales.

**5 Common Blue**, *Polyommatus icarus*. The most abundant and widely distributed of the European Blues and flies in all kinds of open country, up to 1,800m in the mountains and north to Lapland; it extends to Asia and all over the British Isles. The male is always clear blue on the upperside, but the female varies greatly in the proportion of blue to brown. Vetches and clovers are the main food plants.

## LYCAENIDAE

**1 Mazarine Blue**, *Cyaniris semiargus*. Well distributed in Europe and temperate Asia, and it existed in south-western England until the 1870s. In Europe it is most frequent in alpine meadows at moderate altitudes, and it has the habit, so often seen in tropical butterflies, of congregating around puddles and muddy places. The male is uniform blue, the female uniform brown, both with white fringes, and there is very little variation.

**2 Chalk-hill Blue**, *Lysandra coridon*. A butterfly of chalk and limestone hills; it is confined to Europe, including the chalk areas of southern England. The larva feeds on vetches and trefoils and is carried by ants to places near their nest where the food plant grows. Many Lycaenid larvae are attended by ants, and the attraction is in a gland on the back of the larva from which minute drops of a sweet liquid are secreted, which is relished by the ants. The association is comparable with that of ants and aphids.

1 ♂

1 ♀

1

1

**3 Adonis Blue**, *Lysandra bellargus*. The upperside of the male is brilliant shining blue and it is perhaps the most vividly coloured European butterfly. The species inhabits chalky regions where it often flies together with the Chalk-hill Blue, but it has a wider distribution, extending east to Iran. It is found on the chalk of southern England but is now rather local and rare. The butterfly appears in June and again in September; the Chalk-hill Blue has only one brood in the year.

## SKIPPERS

Skippers or Hesperiidae (pp. 66–9) are not closely allied to the other butterflies: the head is wide and the thorax thick and muscular; the larvae live and pupate in silk-spun shelters made of leaves.

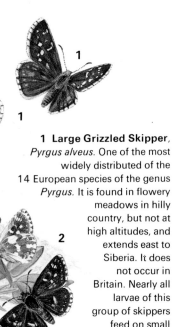

**1 Large Grizzled Skipper**, *Pyrgus alveus*. One of the most widely distributed of the 14 European species of the genus *Pyrgus*. It is found in flowery meadows in hilly country, but not at high altitudes, and extends east to Siberia. It does not occur in Britain. Nearly all larvae of this group of skippers feed on small plants of the family Rosaceae.

**2 Red-underwing Skipper**, *Spialia sertorius*. Owes its name to the brick-red ground colour of the hind-wing underside. It inhabits rough hilly country in central and southern Europe and eastward into Asia.

**3 Grizzled Skipper**, *Pyrgus malvae*. The only species of *Pyrgus* that extends to Britain, where it is confined to southern and central England. It is found all over Europe except in the far north. Its habitat is open flowery places and especially clearings in woods. Food plants include mallow, cinquefoils and wild strawberry.

**4 Mallow Skipper**, *Carcharodus alceae*. Occurs all over central and southern Europe, mostly in hilly country at fairly low altitudes. It may complete four broods during the spring and summer.

**5 Chequered Skipper**, *Carterocephalus palaemon*. A distinctive species found usually in open woodland and ranging from Europe east to Japan. In Britain it is extremely local in the English Midlands, with a totally separate population (more heavily marked with dark brown) in western Scotland. The larva feeds on grasses.

**6 Dingy Skipper**, *Erynnis tages*. A drab little butterfly, widespread from Europe to eastern Asia; it has a preference for chalky soils. In Britain it is common in England, local in Scotland and Ireland. Small leguminous plants provide the larval food.

67

## SKIPPERS

**1 Large Chequered Skipper**, *Heteropterus morpheus*. A quite unmistakable butterfly with its boldly marked underside hind-wings. The female has the fringes chequered black and white and larger yellow markings on the upperside. This is a woodland butterfly occurring locally in widely separated colonies, becoming more frequent in eastern Europe and ranging far into Asia; it has been recorded in Jersey. The larva feeds on grasses.

**2 Silver-spotted Skipper**, *Hesperia comma*. Distinguished by the clear silvery spots on the underside and the distinct markings towards the tip of the upperside fore-wing. It is found in grassy places on chalky soil and is widely distributed in Europe and Asia, extending to western North America. It occurs locally on the chalk of southern England. In this species and nos 3, 4 and 5 the males have a 'sex-brand' of black scent-scales on the fore-wing.

**3 Large Skipper**, *Ochlodes venatus*. Flies in woodland clearings and paths and open grassy places and also on coastal cliffs. The sexes are clearly distinguished by the black sex-brand on the male fore-wing. It is common in Europe and temperate Asia and in Britain it extends to the Scottish Lowlands, but not to Ireland, where the Dingy Skipper is the only Hesperiid butterfly.

**4 Small Skipper**, *Thymelicus sylvestris.* Ranges all over western Europe to Iran but does not extend far to the north, and in Britain does not reach Scotland. It flies in a variety of habitats, rough fields, paths in woods and coastal dunes.

**5 Essex Skipper**, *Thymelicus lineola.* Very like the Small Skipper. The most reliable character for separating them lies in the colour of the tip of the antenna: this is all black in the Essex Skipper but broadly orange coloured below in the Small Skipper, *T. lineola* has a similar distribution in Europe and is found in England mainly near the coast in the south-east.

## HAWK-MOTHS

The hawk-moths or Sphingidae (pp. 70–9) are a very distinct group of moths with thick bodies, long fore-wings and powerful flight. Their larvae almost always have a stiff horn or spine at the hinder end.

**1 Lime Hawk-moth**, *Mimas tiliae*. The most variable of the European hawk-moths. The ground colour ranges from buff to shades of green or brick-red, and the dark markings on the fore-wing may be divided (as in the one shown here), or may form an uninterrupted bar across the wing. The larva feeds usually on lime or linden, but is also found on elm. The moth is often seen in towns where limes are planted; it is common in England and extends from central Europe to Japan.

**2 Eyed Hawk-moth**, *Smerinthus ocellata*. Named on account of the beautiful eye-like markings on the hind-wings. When the moth is at rest they are concealed, but if it is molested it advances the fore-wings and displays the eye-marks, producing an effect rather like the face of a cat or an owl, which can scare predatory birds away. Sallow, willow and apple are the most usual larval food plants and the Eyed Hawk is widespread in Britain and Europe and in temperate Asia.

**3 Poplar Hawk-moth**, *Laothoe populi*. The commonest of the European Hawk-moths and extends north to Lapland and in Britain to the Orkney Islands. Its resting posture with the hind-wings advanced in front of the fore-wings, is characteristic. Poplar is the most usual food plant.

# HAWK-MOTHS

**1 Death's-head Hawk-moth**, *Acherontia atropos*. A large heavily built moth with markings on the thorax that somewhat resemble a skull. If molested the moth makes a squeaking noise by forcing air through the proboscis. These characteristics, its large size and its irregular occurrence have led to its being regarded in Europe with superstitious fear as a portent of ill-luck or disaster. The proboscis is short and stiff and the moth never feeds at flowers, but is known to enter bee-hives and pierce the combs to suck the honey. The larva feeds on the leaves of potato and other plants of the family Solenaceae. Like many hawk-moth larvae it buries itself in the earth to pupate. The home of the Death's-head is in Africa and western Asia, whence it migrates northward into Europe and occasionally reaches Britain. In certain years migrations take place, the most recent being in 1956. Moths arriving in Britain early in the year lay eggs in potato fields, though not many larvae survive modern chemical pesticides; the pupae require artificial heat to produce moths.

**2 Pine Hawk-moth**, *Hyloicus pinastri*. Sometimes a pest in the pine forests of Europe and northern Asia as the Scots pine is the common food plant. The moth's colour and pattern provide good camouflage when it is at rest on pine trunks and the caterpillar is marked with longitudinal dark green and pale coloured stripes to match the clusters of pine needles among which it feeds. At night the moth may be seen hovering in front of flowers, such as honeysuckle, to feed. In Britain the Pine-hawk is confined to eastern and south-eastern England where it has become more common in recent years due to extensive planting of pines.

**1 Convolvulus Hawk-moth**, *Agrius convolvuli.* Very widely distributed in the temperate and tropical regions of the Old World, due to its exceptionally powerful flight and its urge to migrate, which brings it to northern Europe, including the British Isles, as a regular migrant. This big, grey moth feeds by hovering in front of flowers such as honeysuckle. Its proboscis is of extraordinary length, sometimes nearly 10cm long, making it the only moth capable of reaching the nectar of the tobacco plant (*Nicotiana*), to which it is strongly attracted. The larva feeds on species of *Convolvulus*, however it is rarely found in Britain, and the moth cannot overwinter in this country in any of its stages. The pupa, which is enclosed in a cell underground, is peculiar in having a coiled tubular projection in which the proboscis is contained.

**2 Privet Hawk-moth**, *Sphinx ligustri*. The largest hawk-moth resident in Britain and northern Europe. Its Latin name, like that of the family Sphingidae, was prompted by the reared-up attitude of the big green caterpillars, which reminded the early entomologists of the Egyptian Sphinx. The larva feeds mostly on privet and lilac, and the species must depend greatly on the cultivation of these for its existence. Certainly the larvae are most often found on bushes growing in gardens. It is also known to feed on ash, honeysuckle and holly. The moth is easy to breed, but the larva must be given at least 15cm depth of earth for pupation. The Privet Hawk is fairly common in southern England and extends from Europe eastward through Asia to Japan.

**1 Oleander Hawk-moth**, *Daphnis nerii*. The most beautiful of the large hawk-moths. Its home is in Africa and south-western Asia whence it extends into southern Europe migrating northward in small numbers. In Britain it is one of the rarest immigrant species; the year 1953 was exceptional producing 13 records of the moth. The larva feeds on oleander, grape-vine and the lesser periwinkle.

**2 Proserpine Hawk-moth**, *Proserpinus proserpina*. Commemorates the mythical daughter of Ceres, Proserpina, who had to live in Hades for half of every year. In Europe it has a southern distribution, which includes Spain and Portugal. The larva feeds on plants of the willowherb and loostrife families, and is exceptional among the Sphingidae in lacking a horn on the tail.

**3 Spurge Hawk-moth**, *Hyles euphorbiae*. Widely distributed in Europe but has never established itself in Britain, though it is occasionally recorded as an immigrant. It extends eastward to Iran and north-western India. The brightly coloured and patterned larva feeds on cypress spurge, sea spurge and other species of *Euphorbia*. Many of these plants have an acrid and poisonous sap and it is possible that this makes the larvae inedible to birds, and that their lurid coloration is a warning. The species has been introduced into western Canada in an attempt to control some *Euphorbia* species that have become established as weeds.

**4 Silver-striped Hawk-moth**, *Hippotion celerio*. A tropical and subtropical moth found all over the warmer parts of the Old World and extending to Australia. It lives along the Mediterranean coast and occasionally flies north as far as Britain. Grape-vine and Virginia creeper are recorded as food plants.

1

2

2

**1 Broad-bordered Bee Hawk-moth**, *Hemaris fuciformis*. One of a pair of similar European species, the other being the Narrow-bordered Bee Hawk-moth, *Hemaris tityus*; their names indicate the difference between them. Both are day-flying moths which feed by hovering at flowers, and both have the central area of the wings transparent. They certainly bear some resemblance to bumblebees and are believed to derive protection from predatory birds which avoid attacking bees and are deceived by the moths' mimicry. The larval food plant of the Broad-bordered is honeysuckle, that of the Narrow-bordered wild scabious. Both are well distributed in Europe and western Asia and occur in Britain; only the Narrow-bordered extends to Ireland.

**2 Elephant Hawk-moth**, *Deilephila elpenor*. A beautiful insect and not at all rare. It flies at night and feeds from flowers, especially honeysuckle. Its name is derived from the ability of the larva to extend its foremost segments in a way suggestive of an elephant's trunk. When these segments are retracted the eye-spots on the larva produce a resemblance to a snake which may have some defensive value. (The larva of the Silver-striped Hawk (p. 77) is endowed in a similar way.) The main food plants of the Elephant Hawk are willowherb and bedstraw.

**3 Hummingbird Hawk-moth**, *Macroglossum stellatarum*. A very pretty, attractive insect which flies by day. It hovers at flowers to feed and looks remarkably like a hummingbird. It is common all over Europe and temperate Asia and is a frequent immigrant to the British Isles. In warm-temperate climates it hibernates as a moth and may occasionally do so in south-west England. The larval food plant is bedstraw.

## NOTODONTIDAE

Many moths of this family (pp. 80–3) are characterised by unusual larvae.

**1 Sallow Kitten**, *Furcula furcula*. One of a number of smaller relatives of the Puss Moss which are appropriately called 'kittens'. Its early stages are similar to those of the Puss Moss and it is found in Europe including Britain.

**2 Puss Moth**, *Cerura vinula*. Extends from western Europe and Britain to Japan. The larva is noted for two long 'tails' from each of which it extends a thread-like organ if molested; when fully grown it faces its enemy to display an alarming 'face' with two black spots for eyes. It feeds on poplar and willow and pupates in a cocoon which few enemies can penetrate.

**3 Lobster Moth**, *Stauropus fagi*. Named after the extraordinary appearance of its mature larva; when newly hatched they resemble ants. The larval food includes beech, birch and oak and the species extends from Europe to Japan; it is also found in southern England and Ireland.

## NOTODONTIDAE

### Prominent Moths

The so-called 'prominent moths' are characterised by a tuft of scales on the hind-margin of each fore-wing, the two forming a distinct projection when the moth is at rest.

**1 Three-humped Prominent**, *Tritophia tritophus*. Ranges widely throughout Europe, Asia and North America, but not established in Britain, where it occurs only as a rare immigrant. The larva feeds on poplar and has three prominent humps on the back and another over the tail.

**2 Pebble Prominent**, *Eligmodonta ziczac*. Rather like the Three-humped, but the pattern on the fore-wings is brighter and better defined. The larva is similarly shaped to that of the other species but has only two humps on the back. It feeds on sallow, willow and poplar and over-winters, as do most of the Notodontidae, as a pupa. Its distribution in Europe and Asia includes the whole of the British Isles.

**3 Swallow Prominent**, *Pheosia tremula*. A pretty moth with beautifully shaded brown and white fore-wings. Its larva is more normal in appearance than that of the other prominents, but does have a distinct hump over the tail. The species is widespread in Europe and Britain.

**4 Buff-tip**, *Phalera bucephala*. A distinctive species which, when at rest, simulates a piece of broken twig; at each end of the moth there is an area of yellow, representing the bare wood, and between these the silvery-grey fore-wings look like dead bark. The hairy yellow larvae feed on various deciduous trees and when young live crowded together; they are probably distasteful to birds. The species is common in Europe and temperate Asia and all over the British Isles.

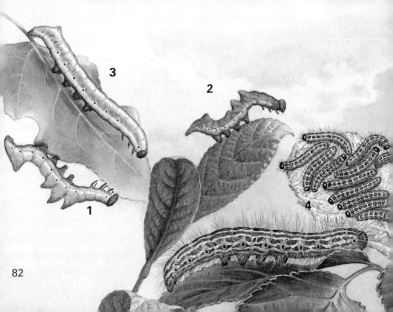

The moths described below belong to the families Drepanidae, Thyatiridae and Thaumatopoeidae respectively.

**1 Pebble Hooktip**, *Drepana falcataria.* One of a group of woodland moths which have the fore-wings hooked like a sickle. Their larvae lack the hindmost prolegs and the body ends in a point. The food plant of this species is birch and it is fairly common in central and northern Europe and all over the British Isles.

**2 Peach Blossom**, *Thyatira batis* and **3 Buff Arches**, *Habrosyne pyritioides.* Two closely allied, very beautiful moths whose larvae feed on bramble. They are found in woodland in Europe and Britain.

**4 Oak Processionary Moth**, *Thaumetopoea processionea.* One of the three European species of this genus and has oak as its main food plant. The Pine Processionary Moth (*T. pityocampa*) feeds on pine and a third, more northern species, *T. pinivora*, is also a pine feeder. None of them are found in the British Isles. The moths are unremarkable brownish insects, but the larvae have extraordinary habits. They live in communal webs spun among the branches and make excursions during the day to feed on the leaves of neighbouring trees. In doing so they crawl head to tail in long processions, laying down a track of silk as they go which guides them back to the web. The Pine Processionary larvae go in single file, and a column of 300, 12m long, has been observed. The Oak Processionary columns have a single leader, but become broader behind with the larvae marching side by side. It seems that the nose-to-tail contact is necessary as well as the silk trail, to keep the column moving. They hibernate in their webs and the Oak Processionary larvae also pupate in the web. These caterpillars possess irritant hairs that may cause serious urticaria if they are handled.

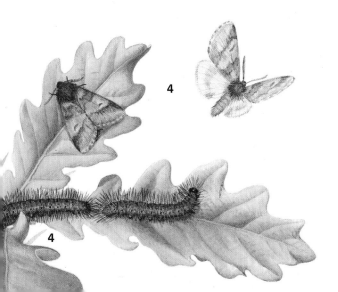

## LYMANTRIDAE

These moths are called the 'despoilers' because the larvae of some of them do serious damage by defoliating trees. The larvae are often brightly coloured and ornamented with tufts of hair and the hairs in some species have irritant properties.

**1 Vapourer Moth**, *Orgyia antiqua*. Found in the temperate regions of Eurasia and North America. The female is stout, heavy and wholly wing-less and does not leave the cocoon which contains her pupa, but waits for the male, which flies by day, to find her and mate with her. The eggs are laid on the cocoon where they overwinter.

**2 Black Arches**, *Lymantria monacha*. Also known as the Nun. Its larva feeds on oak and other trees and is sometimes a minor pest in European orchards, but in Britain it is confined to southern England and is never very common. It extends through temperate Asia to Japan.

**3 Gypsy Moth**, *Lymantria dispar*. A notorious pest of deciduous trees of almost all kinds. The species is native to Europe and all temperate Asia but has been extinct in England since about 1850.

**4 Yellow-tail** or **Gold-tail**, *Euproctis similis*. A silky-white moth widespread in Europe, Asia and the southern parts of Britain. The larva feeds on various trees, most often on hawthorn and sallow, and its bright colours advertise the irritant and poisonous nature of the hairs on its body.

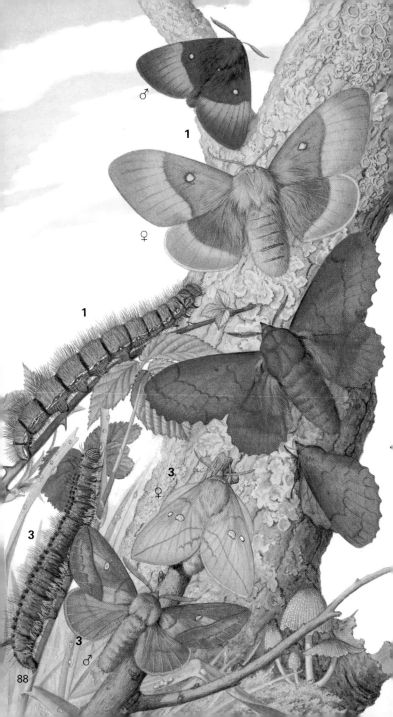

## LASIOCAMPIDAE

The members of the family Lasiocampidae are mostly big moths, the females larger than the males. The larvae are hairy and spin strong silken cocoons.

**1 Oak Eggar**, *Lasiocampa quercus*. Found throughout the British Isles and most of temperate Eurasia in open woods, on heaths and near the sea shore. The male is active during the day, flying wildly in search of females. The larger, paler coloured female flies at night to distribute her eggs. In southern Britain the life cycle takes a year, but in the northern race, *callunae*, it takes two. Despite the moth's name, bramble, hawthorn and ling are usual food plants.

**2 Lappet Moth**, *Gastropacha quercifolia*. Found across Eurasia to Japan, but in Britain only in the southern half of England. When at rest the moth resembles a bunch of dead leaves. The larva feeds on blackthorn and orchard trees but is never abundant enough to do any harm.

**3 Drinker Moth**, *Philudoria potatoria*. So called from its habit of drinking drops of dew. The larvae can be found after hibernation in damp places such as the banks of ditches, where it feeds on coarse grasses. The species extends all over Eurasia including Britain.

## ENDROMIDAE

**4 Kentish Glory**, *Endromis versicolora*. The only member of this family. It is confined to Europe and is still found in Scotland but disappeared long ago from Kent. The larval food plant is birch.

1 ♂

2 ♂

2

2 ♀

## SATURNIIDAE

This family (pp. 90–2) consists of the emperors and giant silk moths and includes the world's largest and most magnificent Lepidoptera. The great majority are tropical, among them the huge Atlas Moth of south-eastern Asia and the species of *Antheraea* which produce tussore silk. Many species have a clear spot or patch without scales in the middle of each wing. The males are characterised by comb-like antennae with which they can detect the scent of the females at great distances.

**1 Great Peacock Moth**, *Saturnia pyri*. The largest European species of either moth or butterfly, with a wing-span of up to 15cm. It is common in southern Europe and western Asia and has been found as far north as Paris. The larva feeds on sloe, apple, elm and other trees.

**2 Emperor Moth**, *Saturnia pavonia*. The only British member of the Saturniidae and widely distributed in Europe and temperate Asia, usually in moorland areas. The moths are active in April and May when the males fly about by day searching for females which they locate by their characteristic scent; the females fly only at night. The main food plant is heather, but larvae are also found on bramble and other plants.

**3 Tau Emperor**, *Aglia tau*. Unlike most Saturniids this species is very variable. The ground colour is typically tawny yellow, but ranges through brown to almost black, and the eye-spots vary in size. It is an insect of beech woods, though it feeds on birch, lime and oak too, and is found quite far north in Europe but not in Britain.

**1 Spanish Moon Moth**, *Graellsia isabellae*. Ranks among the most striking and beautiful moths in the world and is certainly first among the European Lepidoptera. Confined to western Europe, it seems to have evolved in isolation in the Iberian peninsula and was known only from Spain for many years; it is now known also to occur in the French Alps. The larva feeds on pine and the species overwinters as a pupa.

# ARCTIIDAE

The characteristics of this family (pp. 93–7) are described on p. 94.

**2 Cinnabar Moth**, *Tyria jacobaeae*. A solely European species found in most of Britain. Its larval food is the poisonous weed ragwort and both larva and moth are distasteful to predators.

**3 White Ermine**, *Spilosoma lubricipeda*. A common British and European species that extends east to Japan. It is distasteful to birds and the larva is of the 'woolly bear' type, covered with a dense coat of stiff hairs and avoided by most predators. When searching for a place to pupate the larva runs rapidly over the ground, hence the name *lubricipeda*, 'slippery-foot'. It feeds on plantain, dandelion and other low growing plants.

**4 Crimson Speckled Moth**, *Utetheisa pulchella*. A widespread migrant that occasionally strays to Britain. The larva feeds on the borage family.

## ARCTIIDAE

Many of the Arctiidae (pp. 93–7) are protected as moths by distasteful body fluids and they advertise this fact to predators by displaying bright patterns and colours. Some species produce ultrasonic sounds which warn bats (who hunt by echo-location) to leave them alone. Many of the larvae are densely hairy and some induce urticaria if handled.

**1 Clouded Buff**, *Diacrisia sannio*. Found mostly on heaths. The sexes are very different in appearance and fly at different times, the male by day, the female after dark. It occurs widely in Europe and over the whole of the British Isles. Like those of most of the tiger moths the larva feeds on low-growing plants and overwinters in this stage.

**2 Ruby Tiger**, *Phragmatobia fuliginosa*. Extends all round the temperate regions of Europe, Asia and North America. In Britain the moth varies regionally: specimens from Scotland have the red on the hind-wings much reduced.

**3 Wood Tiger**, *Parasemia plantaginis*. A species of moors, chalk downs and open woodland. It is variable, and in northern latitudes and on mountains males but not females have the ground colour of both fore- and hind-wings white. The larva can be recognised by the colour of the hairs, red-brown at mid body, black in front and behind. It feeds on a variety of low-growing plants and hibernates when half grown.

**4 Purple Tiger**, *Rhyparia purpurata*. An eastern European tiger moth that extends to China and Japan. Within its range it is generally common.

# ARCTIIDAE

**1 Scarlet Tiger**, *Callimorpha dominula*. Found in Europe, including southern England and Wales and in western Asia. It is a variable species; one form, common in southern Europe has yellow hind-wings instead of red, others have the extent of the pale spots on the fore-wings reduced. These forms have been named and shown by breeding experiments to be under simple genetic control. The Scarlet Tiger frequents marshes and river banks with dense vegetation and it flies in sunshine. The larva feeds on comfrey, groundsel, nettle and other plants.

**2 Cream-spot Tiger**, *Arctia villica*. Fairly common in southern England and Wales, mainly near the coast, and in Europe and western Asia. It varies in the relative amounts of black and pale coloration; either may predominate on both fore- and hind-wings. It has distasteful properties and performs a warning display of the yellow hind-wings and red abdomen. The moth flies at night in May and June and the larva feeds near the ground on various plants.

**3 Garden Tiger**, *Arctia caja.* A conspicuous and familiar insect. It is extremely variable and a favourite of both collectors and geneticists, both of whom carry out elaborate breeding programmes to produce varieties as specimens or to demonstrate the mechanism of heredity. In the extreme forms all the wings may be predominantly or wholly dark coloured, or they may be almost entirely white and red. The moth has poisonous properties and when molested displays both the hind-wings and the bright red 'collar' behind the head. The larva will eat leaves of almost any kind and can be reared on a diet of cabbage. The species inhabits temperate Eurasia and North America.

**4 Jersey Tiger**, *Euplagia quadripunctaria.* At the edge of its range in Britain and almost confined to south Devonshire. It is quite common in Europe and is the species seen in huge numbers in the 'Valley of the Butterflies' on the Island of Rhodes. The larva feeds on various low plants and the species extends into western Asia.

## NOCTUIDAE

The Noctuidae (pp. 98–105) is an enormous family of mostly nocturnal moths; nearly all the species that visit the collector's bait of treacle belong to this family. Some of the larvae, the 'army-worms' and 'cut-worms', are serious pests of agriculture and many of the moths have 'ears' on the thorax that detect the ultrasonic squeaks of hunting bats.

**1 Merveille du Jour Moth**, *Dichonia aprilina*. Named 'marvel of the day', but is in fact very seldom seen by daylight. The beautiful pattern on its wings closely simulates the lichen on the trunks and branches in the oak woods in which it lives, making it very difficult to find. Oak is the sole larval food plant and in Britain and northern Europe the moth flies in September and October. It is fairly common wherever there are oak woods.

**2 Scarce Silver Lines**, *Bena prasinana*. Another species that is confined to oak woods. It is less common than the rather similar Green Silver Lines, *Pseudoips fagana*, but is not really scarce. The larva feeds on oak and pupates in a cocoon shaped like an upturned boat, spun on a leaf. The moth is found in Britain and Europe west to Asia Minor. The green colour of this moth and of the Merveille du Jour fades if exposed to light and is destroyed by the vapour of cyanide and ammonia if these are used as killing agents.

**3 Red Swordgrass**, *Xylena vetusta*. A species in which the moth is less brightly coloured than the caterpillar which is a general feeder on low-growing plants. The moth ranges widely in Eurasia and all over the British Isles.

**4 Bulrush Wainscot**, *Nonagria typhae*. One of the largest species of stem-feeding wainscot moths. It lives in ditches and marshes where bulrush or reedmace grows, and its colour and texture closely simulate the withered leaves. The larva lives inside the reedmace stems feeding on the pith and moving from one stem to another as it grows. It pupates head-downwards in the stem. The species is common in Europe and the British Isles.

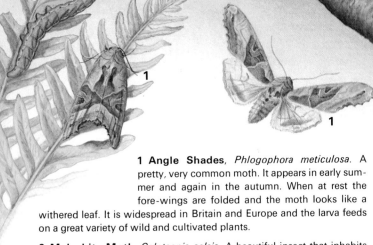

**1 Angle Shades**, *Phlogophora meticulosa*. A pretty, very common moth. It appears in early summer and again in the autumn. When at rest the fore-wings are folded and the moth looks like a withered leaf. It is widespread in Britain and Europe and the larva feeds on a great variety of wild and cultivated plants.

**2 Malachite Moth**, *Calotaenia celsia*. A beautiful insect that inhabits dry open pine forests on sandy soil in eastern Europe and the nearer parts of Asia. Its larva feeds on the roots of grasses.

**3 Old Lady Moth**, *Mormo maura*. This moth received its name long ago when old ladies wore dresses of dark-coloured patterned material. It flies in July and August and the larva starts its life feeding on low-growing plants and then hibernates; when it resumes feeding in the spring it resorts to the young shoots of trees such as sallow and birch. The moth is fairly common in central and southern Europe including Britain; by day it can often be found resting in the dark corners of outhouses.

**4 Large Yellow Underwing**, *Noctua pronuba*. The most abundant of a group of moths with black-bordered yellow hind-wings which are hidden beneath the fore-wings when the moth is at rest. This species is very common in Britain and Europe. The larva feeds on herbaceous plants and is often a pest in gardens.

**5 Broad-bordered Yellow Underwing**, *Noctua fimbriata*. More richly coloured and patterned than the Large Yellow Underwing and is a woodland moth. It is found in Britain and Europe and the larva feeds on birch and other trees.

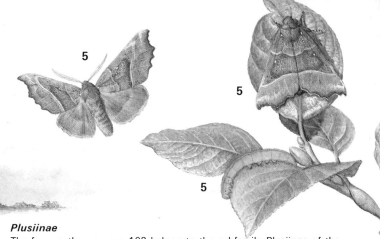

5

5

5

## Plusiinae

The four moths on page 102 belong to the subfamily Plusiinae of the Noctuidae. The name – from the Greek word meaning rich – refers to the gold and silver metallic markings on the wings of most of them. They also have elaborate crests of hair on the thorax and are indeed very pretty insects. They feed at flowers, hovering and probing for the nectar with long probosces. A few of them are habitual migrants. In the larvae the first pair of prolegs is rudimentary or absent.

**1 Burnished Brass**, *Diachrysia chrysitis*. One of the more common Plusiine moths and found wherever nettle and dead nettle, its larval food plants, grow. It occurs all over the British Isles and in Europe and Asia.

**2 Scarce Burnished Brass**, *Diachrysia chryson*. A much less common insect. It is only found where its food plant, hemp agrimony, grows in marshy surroundings, but it does not occur wherever this quite common plant grows. In Britain it is restricted to England but has a wide range in Europe and Asia.

**3 Golden Plusia**, *Polychrisia moneta*. First appeared in Britain in 1890 and has since spread through England to southern Scotland. In this country it seems to be entirely dependent on the cultivated varieties of monkshood and larkspur, which are the larval food plants.

**4 Silver Y**, *Autographa gamma*. A very abundant species that regularly migrates northwards in Europe in the early summer, sometimes in enormous swarms and multiplies rapidly. It flies by day and by night, and is abundant in Britain in autumn, though it cannot overwinter in the British climate. The larva, which feeds on almost any kind of plant, is shown on page 13. It has a wide range in Eurasia, North Africa and North America.

**5 Herald Moth**, *Scoliopteryx libatrix*. An attractive and unmistakable species found all over temperate Eurasia and North America. It overwinters as a moth, often in barns and outhouses and sometimes in numbers together. The larva feeds on sallow and willow.

### Underwings

These moths of the genus *Catocala* illustrate the principle of 'flash coloration'. Almost all of them have brightly coloured hind-wings which are hidden by the duller fore-wings when the moth is at rest. If the moth is disturbed and takes to flight the sudden flash of bright colour may confuse the predator, and it also makes it difficult to see where the moth settles.

**1 Rosy Underwing**, *Catocala electa*. Widespread from south-central Europe to Japan, but only a few specimens have been taken in Britain. It is distinguished from the more common Red Underwing, *C. nupta*, by the more angular marks on the fore-wing. The larva feeds on willow and poplar.

**2 Clifden Nonpareil**, *Catocala fraxini*. Occurs throughout Eurasia but is a rarity in Britain, though it is known to breed in this country. The larva feeds on poplar.

**3 Orange Underwing**, *Brephos parthenias*. Flies in the sunshine in March and April in birch woods and is very difficult to catch. The larva feeds on birch. It is found in suitable localities in the British Isles and most of temperate Eurasia.

**4 Small Dark Yellow Underwing**, *Anarta cordigera*. Like most other members of the genus *Anarta*, it is almost confined to the sub-arctic regions of the northern hemisphere and to high mountains further south, in Britain to the Scottish highlands. The larva of this species feeds on bearberry (*Arctostaphylos*).

## GEOMETRIDAE

This family (pp. 106–13) comprises the moths with 'looper' caterpillars, having only two pairs of prolegs and described on page 11. The moths are nearly all lightly built with slender bodies and ample wings. All the species on these two pages are found in Britain as well as continental Europe and Asia.

**1 Large Emerald**, *Geometra papilionaria*. A beautiful bright green moth found in woods where its larval food plant, birch, grows. Its colour is liable to fade after death and is affected by most killing agents.

**2 Winter Moth**, *Operophtera brumata*. Very abundant and a notorious pest in orchards, though the larva feeds on many other kinds of trees as well; it pupates in the earth. The moth appears during the winter, but only the male flies; the female is wingless and crawls up the trunks of the trees to lay her eggs on the twigs.

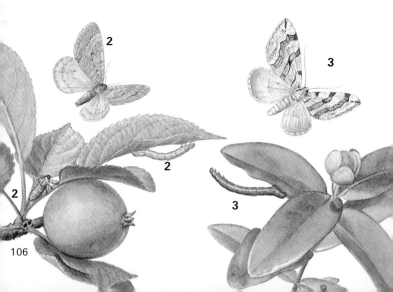

**3 Treble Bar**, *Aplocera plagiata.*
Found in rough open countryside
such as sand dunes or chalk
downs where its food plant,
St John's wort, grows.

**4 Marsh Carpet**, *Perizoma
sagittata.* Its larva feeds on the
flowers and seeds of meadow rue
(*Thalictrum*), a plant of marshy
places. In Britain the moth is
found only in the Fens.

**5 Yellow Shell**,
*Camptogramma bilineata.* A
pretty and very common moth
found in a variety of habitats. It
varies regionally, even within the
British Isles. The larva feeds on
various low-growing plants.

**6 Blood-vein**, *Timandra
griseata.* Another common and
pretty moth characterised by a
red line which runs unbroken
across fore- and hind-wings
when the moth is at rest, possibly
simulating the mid-rib of a leaf.
The larva feeds on docks and
other low-growing plants.

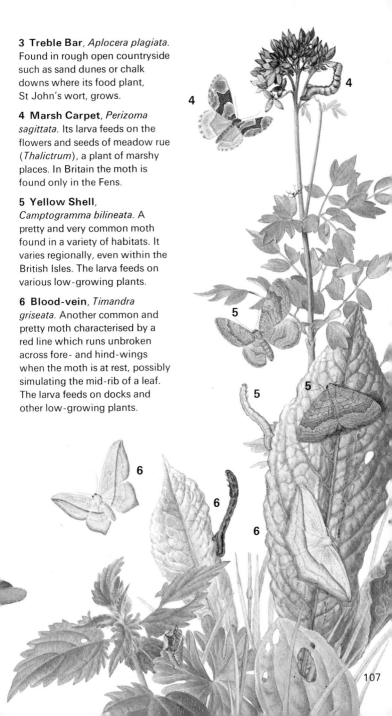

GEOMETRIDAE

1

2

3

♂ 4 ♀

5

5

108

**1 Argent and Sable**, *Rheumaptera hastata*. A widely distributed moth occurring in North America as well as Eurasia. In Britain there are two subspecies, a southern one, illustrated here, whose larval food plant is birch, and a smaller, darker Scottish form which feeds on sweet-gale and bilberry. The moth flies in sunshine from May to July.

**2 Arichanna melanaria**. Found in eastern Europe on wet moors and bogs; its food plant is northern bilberry (*Vaccinium uliginosum*).

**3 Magpie Moth** or **Currant Moth**, *Abraxas grossulariata*. A common species which exhibits extreme variations both in its black and yellow markings and in the ground colour. A number of genetically controlled varieties have been distinguished and have been the subject of research on heredity. This species is an excellent example of a moth which is protected throughout its life by distasteful qualities and warning colours. A British entomologist has experimentally chewed up the larva, pupa and moth and has recorded that all have a bitter taste. The coloration of the larva, like that of the moth, is spotted black, white and yellow and the pupa is black with conspicuous yellow rings. Moth pupae are almost always concealed in cocoons, but that of the Magpie is enclosed in a slight web among the food plant (currant and gooseberry bushes, blackthorn, garden 'euonymus' or heather) and is clearly visible. It is widespread all over Europe and temperate Asia.

**4 Lilac Beauty**, *Apeira syringaria*. A pretty insect which folds its fore-wings in a curious way when at rest. The larva is conspicuously humped and feeds on honeysuckle. It is found in England and Ireland and extends from Europe to Japan.

**5 Large Thorn**, *Ennomos autumnaria*. A handsome moth found right across Eurasia but in Britain only in south-eastern England, where it flies in September. The larva feeds on hawthorn, birch and other trees.

1 ♂

1 ♀

2

3

4

5

The first four figures on the right illustrate the tendency of Geometrid larvae to camouflage themselves by simulating twigs; they are often called 'stick caterpillars'. This stick-like form and colour is combined with a resting posture in which it holds on with the two pairs of prolegs and keeps its body stiff and straight at an angle to the twig it is holding. The resemblance is often enhanced by humps or tubercles which have the appearance of buds on the twig.

**1 Orange Moth**, *Angerona prunaria*. The male and female are differently coloured, the male orange and the female yellow. In the rather uncommon form *corylaria* the orange and the yellow are restricted to the central areas of the fore- and hind-wings, which are otherwise brown. The moths fly in open woodland and are often active early in the evening long before dark. The larvae feed on many kinds of trees and they hibernate through the winter. The moth is not uncommon in well-wooded areas in the British Isles and across Eurasia.

**2 Swallow-tail Moth**, *Ourapteryx sambucaria*. A common species in the British Isles and most of temperate Eurasia. It is the only member of the genus and is quite unmistakable. The name indicates an association with elder, but the larva feeds on other trees too, also on ivy.

**3 Scorched Wing**, *Plagodis dolabraria*. Has a pattern of fine lines crossing the wing that does give it a 'scorched' appearance. It is a woodland moth, attached to oak, birch and beech, the larval food plants, and is found in the British Isles and all over Eurasia. It flies in May and June.

**4 Brimstone Moth**, *Opisthograptis luteolata*. An extremely common, pretty yellow moth which flies throughout the summer all over Europe and in temperate Asia. The larva feeds on blackthorn and hawthorn.

**5 Speckled Yellow**, *Pseudopanthera macularia*. An attractive little moth that flies in open mixed woodland in sunshine in late May and June. It is very variable and specimens with the dark spots heavily predominant or small and widely spaced are not uncommon. The larva feeds on wood sage and other plants of the family Labiatae.

1

2

3

4

5

**1 Peppered Moth**, *Biston betularia*. Exists in Britain and Europe in two distinct forms: a speckled one **(1)** and a black one **(1a)**. The latter has only been established since the middle of the last century; it appeared by selection and evolution, in response to the blackening of the bark of trees in industrial areas. On trees covered with lichen it is the speckled form that is well concealed and so protected from birds. Both still exist, each predominating in the habitat for which it is adapted. The larva feeds on most deciduous trees and shrubs.

**2 Oak Beauty**, *Biston strataria*. Flies early in the year when the trees are bare, and is wonderfully concealed when at rest on lichened oak trunks. The larva feeds on oak and other trees and the moth extends from Britain and Europe to Asia Minor.

**3 Bordered White** or **Pine Looper**, *Bupalus piniaria*. Attached to conifer trees and the larva is often a serious pest in plantations. The sexes are distinct and there are two forms of the moth, a northern one in which the ground colour of the male is white and a southern in which it is yellow. They occur in northern and southern Britain respectively and are widespread in Eurasia.

**4 Mottled Umber**, *Erannis defoliaria*. A widespread species with a wingless female looking like a six-legged spider; the moths are active in winter. The larva feeds on most deciduous trees.

**5 Engrailed**, *Ectropis bistortata*. Flies in March and April and in a second brood in July. The larva feeds on various trees and the species is found throughout Britain and Europe to eastern Asia.

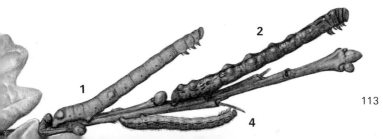

## SWIFT MOTHS

The swift moths, family Hepialidae, are the most primitive of the larger moths. Their larvae either burrow into wood or feed on roots underground.

**1 Ghost Swift**, *Hepialus humuli*. The sexes are very distinct and the name refers to the glistening white male which is conspicuous as it flies at dusk. The males fly and hover over grassy places and the females see them and fly to them; in most Lepidoptera it is the male that seeks out the female. The larva feeds on roots underground and the moth extends from Britain and Europe to western Asia.

**2 Orange Swift**, *Hepialus sylvina*. The sexes differ to a lesser degree; the female is larger than the male and has brown rather than orange forewings. The larva feeds on roots, including those of bracken, and the moth is most often seen where bracken grows. A well distributed species but not found in Ireland.

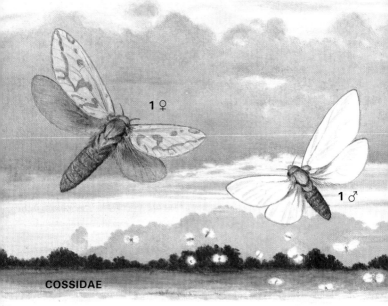

**1 ♀**

**1 ♂**

## COSSIDAE

This family of moths burrows and feeds as larvae in wood and plant stems.

**3 Leopard Moth**, *Zeuzera pyrina*. A conspicuous insect whose larva can do much damage to the wood of trees. Widespread in Eurasia and North America.

**4 Goat Moth**, *Cossus cossus*. So called from the unpleasant smell of its larva. It lives in the wood of trees for several years before it is fully grown. Found in Britain and Europe and east to central Asia.

**1 ♂**

♀ 3

♂

2 ♀

2

4

4

4

3

3

4

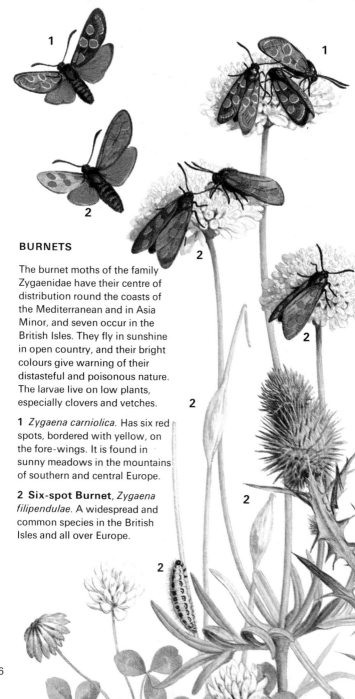

## BURNETS

The burnet moths of the family Zygaenidae have their centre of distribution round the coasts of the Mediterranean and in Asia Minor, and seven occur in the British Isles. They fly in sunshine in open country, and their bright colours give warning of their distasteful and poisonous nature. The larvae live on low plants, especially clovers and vetches.

**1** *Zygaena carniolica*. Has six red spots, bordered with yellow, on the fore-wings. It is found in sunny meadows in the mountains of southern and central Europe.

**2 Six-spot Burnet**, *Zygaena filipendulae*. A widespread and common species in the British Isles and all over Europe.

## CTENUCHIDAE

The moths of this family resemble burnets and are also day fliers, though they are more closely related to the Arctiidae (pp. 93–7) and have similar hairy larvae. Nearly all are tropical but a few occur in Europe.

**3** *Syntomis phegea.* The commonest European Ctenuchid, it is found in southern Europe and Asia and once or twice recorded in England.

## CLEARWINGS

These moths, also known as the Sesiidae, form a distinct family with partly transparent wings so that the moths look like wasps and probably derive protection from the resemblance. The larvae feed in tree-trunks, stems and roots. Both moths described below are found in Europe including Britain.

**4 Hornet Clearwing Moth**, *Sesia apiformis.* Feeds as a larva in poplar trunks.

**5 Lunar Hornet Moth**, *Sphecia bembeciformis.* Feeds on the wood of sallow or poplar.

# Adaptations against Predators

Many of the Lepidoptera are large insects and are often distinctively coloured and patterned. These characteristics make them not only ideal subjects for collectors and photographers but excellent material for illustrating some aspects of biology. One of these is, of course, genetics, because minor variations between individuals are so readily recognisable. (See *Butterflies* and *Moths*, both by E. B. Ford in Collins's New Naturalist series.) Another is the study of the adaptations they have evolved against predators that hunt by sight.

Butterflies and moths have numerous enemies that use them as food and in doing so help to prevent them from multiplying to excess. The most effective of these are other insects, especially the flies and ichneumon wasps whose larvae live on the early stages of Lepidoptera as internal parasites: but it is difficult for us to appreciate the sensations and impulses which make up the hunting behaviour of an ichneumon wasp, and protective measures directed against them will be equally difficult to recognise. Birds, which are also important enemies of Lepidoptera, have sight as their dominant sense, just as we do, and they are vertebrate animals with patterns of behaviour near enough to our own to be readily understood. We will confine ourselves accordingly to the easily recognisable adaptations of butterflies and moths against hunting birds.

Like all insects they are small animals and they are unarmed and defenceless and can never fight off an attack by a bird. Most Lepidoptera escape from these formidable enemies by hiding from them, at any rate when they are at rest, as most of them are throughout the day. Great numbers do this by sheltering in ill-lighted situations: the Old Lady Moth (p. 101) is often found by day in sheds and outhouses and no doubt also shelters in hollow trees and caves. Many of the Noctuid moths (pp. 98–101) that fly in hundreds at night and simply disappear during the day are resting on the ground under thick herbage or brambie or heather. Motionless and 'asleep' some no doubt fall victims to the round-the-clock hunting of shrews, but few are found by birds. The more lightly built Geometrid moths (pp. 106–114) tend to hide in the same way in ivy and the foliage of trees and bushes, whence they can be disturbed by beating or shaking the branches. The few butterflies that hibernate as imagines hide away above ground level, the Comma and Brimstone in overwintering foliage, the Small Tortoiseshell and Peacock in hollow trees, outhouses and attics.

## Camouflage

Being under cover and so out of sight is their first line of defence, but most of these Lepidoptera that hide by day or through the winter will be found to be coloured and patterned to harmonise more or less closely with their surroundings. They are in fact camouflaged or, to use the more formal biological term, they are examples of cryptic adaptation. This form of protection is more highly developed among insects than anywhere else in the animal kingdom, and for diversity and perfection of cryptic devices the Lepidoptera are outstanding among the insects. It is worth making a short digression to discuss the reason for this.

The Lepidoptera, in the course of their evolution, have arrived at a means for the colouring and patterning of their wings which puts them far in advance of any other insects in this respect. This is of course the mosaic effect of closely arranged overlapping coloured scales described on page 8. The endless diversity and beauty of butterflies and moths is to us the most obvious product of this feature, but for the insects themselves it provides a means of displaying visual signals for a variety of purposes. One of these is intraspecific recognition, especially between day-flying males and females, but that is outside our present scope. The other important type of signal is protective, either a cloak of invisibility or, as we shall see later, a lurid warning that the insect is inedible. In the first of these roles the endless diversity of wing pattern and colour is an important ingredient of success. However perfect an example of camouflage may be, extensive repetition of it will lead to sharp-eyed birds acquiring what is known as a 'searching image', in more homely speech they learn what to look for and get their eye in.

In most groups of insects the posture of the wings when at rest is fairly constant. Almost all bees and wasps lay the wings back over the body with only minor variations in their position. This is also true of flies, and most other orders of winged insects have a posture that is characteristic of all of them. But is is far from true of the moths and butterflies, which may hold them up and closed over the back extended on either side with all four wings exposed or only the fore-wings, or laid back in various ways, and the wings may be kept flat or be rolled or folded. This diversity of resting wing posture has surely been evolved to enable the Lepidoptera to exploit to the full the adaptive potential of their wing coloration. The arrangement of the wings almost always plays a part in cases of particular resemblance to twigs, leaves and the like, and butterflies can present two wholly distinct images to the world. With the wings spread as it feeds on a flower a Comma butterfly (p.35) is conspicuous and easily recognised by others of its species, but it is alert and wary too. When it settles down in the autumn in a hedge or thicket for the months-long sleep of hibernation the Comma closes its wings so that only the dark brown mottled underside is exposed. Now the colour and irregular outline of the wings perfectly simulate a tattered withered leaf, and the hungry questing birds pass it by.

It is then by a combination of diversity in colour, pattern and resting posture that the butterflies and moths have evolved as champions of natural camouflage.

Leaf simulation is quite general among moths. The Scarce Silver Lines (p. 98) is coloured bright green and no doubt well concealed among live foliage, but green is not a common colour among Lepidoptera and cryptic imitation of brown dead leaves, like that of the Comma, is much more frequent. The Lappet Moth (p. 88) is an interesting example because its peculiar wing posture greatly enhances the deception. All four wings are dark reddish-brown and scalloped at the edges. When at rest the moth holds its fore-wings sloping roof-like over its back, but the hind-wings are held flat and project in front of the fore-wings so that the moth looks like a bunch of dead leaves. The Poplar Hawk-moth (p. 71) also rests with the hind-wings projecting in front of the fore-wings and produces rather the same effect, but with less perfection. The pattern and the peculiar folding of the fore-wings of the Angle Shades (p. 100) combine to produce an effect that is extremely cryptic among fallen dead leaves.

Imitation of twigs is less usual, but the Buff-tip (p. 83) affords a beautiful example of it. The fore-wings are rolled round the body to produce a cylindrical form and except at the tip are grey like dead bark. The wing-tips and the thorax are yellowish-buff to represent the bare wood at each end of the broken twig. If a Buff-tip is disturbed or dislodged it does not attempt to fly but falls to the ground and lies there motionless. On the floor of woodland a broken twig is in no way a noticeable object.

Simulation of bark by moths which rest by day on the trunks and branches of trees is a frequent form of camouflage. In rural areas tree bark is commonly covered with lichen, and beautiful examples of lichen simulation are encountered among woodland moths. That of the Merveille du Jour (p. 98) is an outstanding case, and the Oak Beauty (p. 112) affords another. This species flies in oak woods in early spring when the trees are bare of leaves and only their trunks and branches are available as resting places. The Pine Hawk-moth (p. 73) is an example of a moth that simulates bark without any lichen covering, the bark of pine trees which is usually bare.

The appearance of larvae and pupae has of course nothing to do with wing patterns, but the early stages of butterflies and moths show quite a range of adaptation. Many larvae, including those of some butterflies, hide at ground level by day and emerge to feed only at night. Of those that spend the whole of their time among foliage a great number are wholly or partly green, but one form of elaborate cryptic disguise is found among them, and that is imitation of twigs, for which they are well fitted by their elongate cylindrical shape. This is found mainly among the larvae of the Geometrid moths and some examples are shown on page 111. Geometrid larvae lack all but the two hindmost pairs of prolegs, and they use these to hold tightly to a twig when at rest, extending the body stiff and straight at the sort of angle at which a branching twig would normally grow. The colour and texture of the caterpillar's body frequently simulates very exactly the twig of the bush or tree which is the usual food plant, and some of these larvae have to be touched by a non-entomological observer to establish their animal rather than vegetable nature.

Nearly all moth pupae are hidden in cocoons, often underground. Cryptic adaptation of the cocoon is sometimes encountered, that of the Puss Moth is described on page 80. Not only is it almost invisible against the bark on which it is spun, but it is also too hard to be broken into by small birds. In the tropics there are examples of elaborate twig and leaf simulation by the exposed pupae of butterflies, but in Europe all that can be said of them is that they are diverse and achieve inconspicuousness by not looking like anything edible rather than by imitating the appearance of inedible objects.

## Safety in Nastiness

We will turn away now from insects which do all they can to escape notice to those which seem to aim in just the opposite direction. The burnet moths (p. 116) are day fliers in open country. They sit on flowers and fly slowly about in the sunshine, and nearly all have bright red hind-wings and spots or blotches of the same colour on the fore-wings. So far from concealing themselves they seem to do all they can to be conspicuous. Birds, however, disregard them, and with good reason: their body fluids contain histamine and hydrocyanic acid and are intensely distasteful and poisonous, and an inexperienced bird that pecks one suffers acute discomfort. The bird is none the worse for the encounter but will avoid burnet moths in future if it recognises them, and this is why, besides being unpleasant tasting, they must be unmistakable even if this means being conspicuous. A pattern of bright red serves this purpose because birds have good colour vision and are known to distinguish red particularly well. The sacrifice of the relatively few moths that are attacked benefits the species as a whole and natural selection promotes adaptations of this kind without any regard for the 'rights' of the individuals which are killed or mutilated.

The display of colours or form or behaviour as a warning of unpleasant or dangerous properties is called aposematic adaptation. The Lepidoptera are as adept at this as they are at camouflage. It is not a feature of European butterflies, but a number of cases, in addition to the burnets, are encountered among the moths. One is the Cinnabar Moth (p. 93) which has a colour scheme similar to that of the burnets; the larva of the Cinnabar also has an aposematic pattern of black and orange stripes. The unpleasant qualities in this case are derived partly from the poisonous food plant, ragwort, and partly from the moth's own metabolism. The Magpie Moth is also known to be distasteful in all its stages and has aposematically coloured larva, pupa and imago as described on page 108. Finally, many of the tiger moths of the family Arctiidae have conspicuous warning coloration and sometimes a special pattern of behaviour that goes with it (pp. 94–97).

## Mimicry

Other insects, of course, are aposematically protected, among them the wasps, which both sting severely and taste unpleasant; their black and yellow stripes give warning of this. The resemblance of the Hornet Clearing Moth (p. 117) to a wasp or hornet is believed to be no coincidence but a case of protective mimicry. This is the term used to describe the situation when a harmless and palatable 'mimic' evolves in such a way as to resemble an aposematically protected 'model'. In the tropics several families of butterflies, the most widespread being the Danaidae, are inedible due to properties derived from poisonous food plants. Edible butterflies of other families mimic these conspicuous models with such exactitude that there can be no reasonable doubt of the reality of the mimic to model relationship. The well known North American Monarch is one of the few Danaidae butterflies found in temperate regions, and is poisonous and aposematically coloured. It is closely mimicked by the Viceroy, which is a Nymphalid allied to the White Admiral and palatable to birds. Aviary birds which have had the unpleasant experience of trying to eat a Monarch will thereafter refuse both Monarchs and Viceroys. In Europe there are no very convincing examples of mimicry within the order Lepidoptera.

A very different sort of mimicry does have examples in Europe. This is the occurrence of eye-like markings which apparently remind an attacking bird of the face of an owl or a cat or something of the kind strongly enough to persuade it to reconsider attacking a victim. This is most effective if the 'eyes' are suddenly revealed, as in the case of the Eyed Hawk-moth, described on page 71. It is possible that the conspicuous eye-spots of the Emperor Moth (p. 90) are also effective in this way. The eye-like markings on the fore part of the body of the larva of the Elephant (p. 79) and Silver-striped Hawk-moths (p. 77) are also mimetic, giving the larva an appearance like that of a snake. It is true that in temperate Europe snakes are not important enemies of birds, but many of our summer visitors have migrated from the tropics where tree snakes are numerous and often prey on small birds. In tropical countries larvae with eye-spots suggesting snake mimicry are of frequent occurrence.

## Deflection Characters

A rather constant feature of the Satyrid butterflies or 'browns' (pp. 44–55) is the presence of small eye-spots or ocelli near the outer margin of the wings, usually more strongly developed on the underside. These certainly play no part in intimidating predators but are believed to be a protective adaptation nevertheless. When a bird attacks a butterfly it is likely to peck at the most noticeable part of it, and in these Satyrids this is at the same time the least vulnerable part – the wing margins bearing the conspicuous ocelli. In this context they are called 'deflection marks' because they deflect the attack away from the insect's body. Satyrids that have obviously had pieces pecked out of their wings in this way are not infrequently seen flying about none the worse for the encounter.

## Flash coloration

Finally we come to the curious phenomenon of 'flash coloration'. In the yellow-underwing and red-underwing moths (pp. 101, 104) the hindwings are brightly coloured and are suddenly revealed as an 'explosion' of colour when the insect flies, and it disappears with equal suddenness when it alights. Some butterflies with cryptic undersides and bright uppersides produce the same effect, but in the moths the bright colours are never revealed except in flight and cannot be recognitional as they are active only at night. It is supposed that the flash of bright colour confuses a predator that disturbs the insect by day and it also may make it difficult to see where it alights. Flash coloration is also seen in grasshoppers and various other insects, but its effectiveness as a protective device is difficult to demonstrate experimentally. However, no other theory provides any explanation of it at all.

# Index of English Names

# Index of Scientific Names